RECOMBINANT DNA
LABORATORY MANUAL

RECOMBINANT DNA LABORATORY MANUAL

Judith W. Zyskind
Sanford I. Bernstein

San Diego State University
San Diego, California

Academic Press, Inc.
Harcourt Brace Jovanovich, Publishers

San Diego New York Boston Berkeley London Sidney Tokyo Toronto

Academic Press, Inc.
San Diego, California 92101

United Kingdom Edition published by
Academic Press Limited
24–28 Oval Road, London NW1 7DX

Library of Congress Cataloging-in-Publication Data

Zyskind, Judith W.
   Recombinant DNA laboratory manual.
   Includes index.
   1. Recombinant DNA—Laboratory manual.
I. Bernstein, Sanford I. II. Title.
QH442.Z97   1989   574.87'3782   88-7970
ISBN   0-12-784400-7   (alk. paper)

Printed in the United States of America
89 90 91 92   9 8 7 6 5 4 3 2 1

This book is dedicated to our parents,
Max Correy Weaver, Mary Landis Weaver,
Harold Bernstein, and Adele Bernstein.

CONTENTS

Contents

Lab VIII. DNA Gel Blotting, Probe Preparation, Hybridization, and Hybrid Detection 117

Lab IX. Lambda Phage Manipulations 135

A. Phage Plating and Plaque Transfer 135

B. Bacteriophage Lambda Miniprep 144

Contents

This book represents the culmination of more than five years of effort directed at initiating and refining a laboratory course in recombinant DNA techniques. We have approached this course from two different backgrounds: J.W.Z. utilizes bacterial genetics and molecular biology in her research program; S.I.B. is a specialist in eukaryotic gene manipulation. As such, we have attempted to incorporate procedures that are useful to individuals studying either prokaryotes or eukaryotes, keeping in mind that the ability to manipulate *E. coli* and its phages is requisite for any DNA molecular biologist. The experiments have been tested for several semesters in the classroom and are designed to teach the fundamentals of recombinant DNA technology. These techniques include basic microbiological manipulations; extractions of chromosomal, plasmid, M13, and lambda phage DNA; gel electrophoresis; *in vivo* mutagenesis; restriction mapping; isolation and cloning of restriction fragments; DNA sequencing; probe labeling (with both biotin and radioactive precursors); DNA gel blotting; phage plaque blotting; and molecular hybridization. Dr. Douglas Smith has written a chapter on computer analysis of DNA/protein sequences. As computer literacy is an integral part of the modern scientific approach, we include several hands-on computer sessions in our course in which students enter and manipulate sequence information. These computer exercises are specific for our particular program and computer configuration and thus are not included in this manual.

We have designed this manual for those who have a background in molecular biology but have limited experience in manipulating DNA molecules. As such, this book should also prove useful to the advanced undergraduate or beginning graduate student. The techniques described in the laboratory exercises and those in the appendix will allow this manual to be useful beyond the classroom situation. It should be particularly helpful to established investigators who are changing their research focus.

The experiments are designed to be included in a semester-long course that meets twice weekly for three hours per session. As a guide to the instructor, we have included the schedule we utilize for teaching our course. This can, of course, be modified to accommodate the quarter system or laboratory periods of a different length than ours.

The laboratory course (along with a lecture course

we teach) serves as the major element of a Recombinant DNA Certificate program at San Diego State University. We began the program in 1984, and many of our students have gone on to advanced degree programs or careers in the biotechnology industry. In designing such a program, we regard it as essential that both technological and basic research advances be presented to the students. A lecture course is necessary for providing such a background, and we urge prospective instructors to incorporate oral and written student presentations and analysis of journal articles as a major component of such a course.

We are indebted to members of the Biology Department at SDSU and Deans Donald Short and James Neel for helping us to initiate the Recombinant DNA Certificate program. The contributions of several graduate assistants (Teresa Larsen, Elisabeth Roche, Martine Uittenbogaard, Rafael Soto-Gil, Patrick O'Donnell, David Becker, Annabeth Fieck, and Paul Thomas) who aided us in designing and troubleshooting the laboratory experiments have been essential to the success of our course. We also thank Robert Phelps, whose organizational and preparative skills make the laboratory class flow much more easily. We are pleased that International Biotechnology Incorporated (a Kodak company) of New Haven, CT has agreed to sponsor our laboratory course. We are indebted to Matthew Biery, Michael McCaffery, William Kronert, and Kouakou Golly for their help in preparing some of the figures and to Anne Chiaramello and Norbert Hess for help in preparing the protocols in the appendix. Finally, we thank Charles Arthur, who as editor for this book always had encouragement when we most needed it.

Judith W. Zyskind
Sanford I. Bernstein

SCHEDULE OF LABORATORY EXERCISES

Period	Lab I	Lab II	Lab III	Lab IV	Lab V	Lab VI	Appendix
1	(Introduction)						
2	Day 1						
3	Day 2	Day 1					
4		Day 2	Day 1				
5		Day 3	Day 2	Day 1			
6		Day 4		Day 2			
7		Day 5			Day 1		
8					Day 2	Day 1. Part A	
9					Day 3	Day 1. Part B	
10					Day 4	Day 2. Part A	
11					Day 5	Day 2. Part B Day 3	
12						Day 4	
13						Day 5	
14						Day 6	

Schedule of Laboratory Exercises

Period	Lab VI	Lab VII	Lab VIII	Lab IX	Appendix
15	Day 7				Large-Scale Plasmid Prep Day 2
16	Day 8	Day 1			Day 3
17		Day 2			Day 4
18		Day 3			Day 5
19		Day 4			Day 6
20		Day 5		A. Day −1	Day 7
21				A. Day 0	
22			Day 0	A. Day 1 B. Day 1	
23			Day 1	A. Day 2 B. Day 2	
24			Day 2	B. Day 3	
25			Day 3	B. Day 4	
26			Day 4	B. Day 5	
27			Day 5		
28	(Checkout)				

BACTERIAL GROWTH PARAMETERS

It is often important to be able to determine three properties of a bacterial cell culture: its phase of growth, the growth rate, and the cell concentration. Generally, because experiments performed on exponentially growing cells are most readily reproducible, many procedures start with cells in this phase of growth. Also, the success of many DNA isolation procedures depends upon using a given number of cells. This experiment will provide you with the tools for easily determining these parameters for a particular strain of bacteria grown in a specific medium at a certain temperature. The approach is to correlate the number of cells in a growing culture of *Escherichia coli* with the optical density (OD) at 450 nm of that culture and, by following the changes in OD_{450} and cell number during growth, to determine the growth rate. In future experiments, if the strain of *E. coli*, the growth medium, the temperature at which the cells are grown, and the wavelength you use to determine the culture turbidity all remain unchanged, then the growth phase, growth rate, and cell density can be determined by simply following the OD_{450} of the cell culture. If you always start with a culture grown overnight (overnight culture) under the same conditions and inoculate the new culture from the overnight culture using the same dilution factor, then you can predict at approximately what time the culture will be ready for the experiment.

In this experiment, the OD of the bacterial culture is measured in a spectrophotometer. The OD of a culture is primarily the result of light scattering by the bacterial cells, because most bacteria do not absorb visible light. When the OD is plotted on the log scale and time is plotted on the linear scale of semilog paper, it is easy to estimate the time it takes for the OD to double, and because OD is directly proportional to cell number in the exponential growth phase, this is the generation time (*g*) of the culture.

Measuring Bacterial Cell Growth

MATERIALS

❑ Exponential phase culture of *E. coli* strain LE392 *lacY galK2 galT22 metB1 trpR55 supE44 supF58 hsdR514* (r_k^-, m_k^+)

❑ 50 ml of sterile LB medium in a 250-ml culture flask

❑ LB agar plates

❑ Dilution tubes (sterile)

❑ Sterile M9 salts for making dilutions (6 g/l Na_2HPO_4, 3 g/l KH_2PO_4, 0.5 g/l NaCl, 1 g/l NH_4Cl)

❑ Glass petri plates containing 95% ethanol

❑ Bent glass rods for spreading bacteria on plates

❑ 37°C shaker incubator

❑ Spectrophotometer

❑ 37°C air incubator

Note: An overnight culture is made by inoculating a small volume of culture medium (5–10 ml) with cells from a single colony. After the culture has grown overnight, it is then diluted (usually 1:100) into fresh medium and grown to exponential phase.

PROTOCOL

1. Add 1 ml of the exponential phase culture of *E. coli* strain LE392 to 50 ml sterile LB broth. Start timing the age of the culture at this point, and shake the culture at 37°C except when samples are removed for optical density (OD) measurements and viable cell counts.

2. Measure the OD at 450 nm of a 5.0-ml aliquot of the *E. coli* strain LE392 culture at 0 time and at 30, 60, 90, and 120 minutes to determine turbidity.

Note: The starting culture is in exponential phase in order to minimize the length of the lag phase so that the experiment can be accomplished in 3–4 hours. Normally, you would be diluting from an overnight culture if you were in an experimental laboratory.

3. Plate dilutions of the culture as described in Table 1-1 to determine cell number. In plating dilutions, 0.1 ml of the dilution is spread on the agar plate. Thus, for a final dilution of 10^{-3}, you would plate 0.1 ml of the 10^{-2} dilution. Spread the bacteria around the agar plate using a bent glass rod that you have sterilized by dipping in ethanol and flaming. To prevent a fire, never return the glass rod to the ethanol solution until the flame is extinguished.

4. Incubate the plates inverted at 37°C overnight or at room temperature for two days.

Caution: In plating dilutions, sterility must be maintained. For example, if you placed the 5 ml for the OD reading in an unsterile spectrophotometer tube and then removed cells for dilution from this tube, your dilutions would be contaminated.

Table 1-1. Plating Procedure for Determining Cells/ml

Time	Dilutions to make	Final dilution after spreading 0.1 ml of last dilution on plates
0, 30, 60 minutes	10^{-3} (1:100, 1:10)	10^{-4}
	10^{-4} (1:100, 1:10, 1:10)	10^{-5}
	10^{-5} (1:100, 1:10, 1:10, 1:10)	10^{-6}
90, 120 minutes	10^{-4} (1:100, 1:100)	10^{-5}
	10^{-5} (1:100, 1:100, 1:10)	10^{-6}
	10^{-6} (1:100, 1:100, 1:10, 1:10)	10^{-7}

Plotting Cell Growth Data

PROTOCOL

1. Count the colonies on the plates. Plot the following:
   a. OD_{450} (y) vs. time (x) on semilog graph paper;
   b. cells/ml (y) vs. time (x) on semilog graph paper;
   c. $\log_{10}$ cells/ml (y) vs. time (x) on linear graph paper;
   d. OD_{450} (y) vs. cells/ml (x) on linear graph paper. Draw a line through the data points after you calculate the slope (m) using the least-squares equation.

$$m = \frac{\Sigma(x)\Sigma(y) - n\Sigma(xy)}{[\Sigma(x)]^2 - n\Sigma(x^2)} \qquad (1\text{-}1)$$

The intercept, b, is expected to be 0 for the definition of this line, $y = mx + b$; n is the number of data points. In the future, you can use the slope of this line, m, to calculate the number of cells present in your culture from OD_{450} measurements.

Note: There are two possible reasons why your data may not fit a straight line in the plot of OD_{450} vs. cells/ml: (1) the size of the cell changes throughout the cell cycle; cells in the lag and stationary growth phases are smaller than those in the exponential phase. (2) At higher cell densities, a ray of light has an increased probability of being doubly scattered so that the chance of reaching the photodetection system is increased. You may, therefore, want to connect the data points and use this line in the future to determine the cell density from OD measurements rather than using the slope obtained from the least-squares equation.

2. Calculate the time it takes for the cell number to double (doubling or generation time, g) from the following equation using either the cell number at 60 minutes for N_0 and the cell number at 90 minutes for N, or, if these values do not fall on the linear part of the growth curve in the exponential growth phase, use values for N_0 and N that do. See Ingraham *et al.* (1983) for a detailed explanation of the exponential growth phase.

$$\ln N - \ln N_0 = kt \qquad g = 0.693/k \qquad (1\text{-}2)$$

3. Confirm that g is correct from the graphs drawn in steps 1a and 1b. Indicate different growth phases present in graphs a, b, and c as shown in Fig. 1-1.

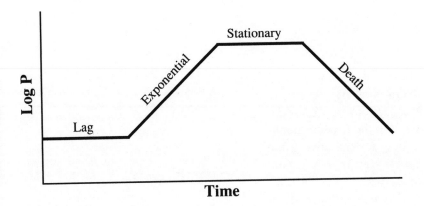

Figure 1-1. Growth curve of a bacterial culture showing lag, exponential, stationary, and death phases. P refers to any parameter of biomass being measured including viable cell number, dry weight, OD_{450}, or some cell component such as DNA, RNA, or protein. The transition from one growth phase to another varies, depending upon the parameter being measured. This is especially true for cell mass and cell number. During the lag phase, the increase in cell mass occurs much earlier than cell division, and in the transition to the stationary phase, the increase in cell mass falls behind the rate of cell division. However, during exponential growth, cell mass and cell number increase at the same rate. These differences should be reflected in the growth curves derived from OD_{450} (a) or cell number (b and c) measurements if you start with an inoculum from an overnight culture and follow the growth of the culture into the stationary phase.

4. Determine how you would use the data from this experiment in the future when a procedure calls for *E. coli* cells in a certain phase of growth at a certain cell density.

ADDENDUM

A. Choosing an *E. coli* strain for cloning

The most important phenotype required for successful cloning of heterologous DNA into *E. coli* cells involves the *Eco*K restriction–modification system encoded on the *E. coli* chromosome. There are three genes in the *Eco*K restriction–modification system: *hsdR*, *hsdM*, and *hsdS*. Mutations in the *hsdR* gene are deficient in the *Eco*K endonuclease activity but still retain the *Eco*K methylase activity giving the phenotype $r^- m^+$. Mutations in either *hsdM* or *hsdS* lead to loss in both the endonuclease and methylase and have the phenotype $r^- m^-$. The optimum phenotype is $r^- m^+$, because DNA isolated from this strain will be methylated at *Eco*K sites and can be introduced into $r^+ m^+$ *E. coli* strains without concern for degradation by the *Eco*K endonuclease. Although heterologous DNA can be successfully cloned in *hsdM* or *hsdS* mutants, DNA isolated from these mutants is not methylated at *Eco*K sites and will be degraded just as foreign DNA will be when introduced into $r^+ m^+$ *E. coli* strains. The optimum mutation, therefore, is in the *hsdR* gene, which is in the strain used in this experiment, LE392. See Raleigh (1987) for a discussion of the interference with cloning caused by the r^+ phenotype in *E. coli*.

Other *E. coli* mutations of interest and concern when cloning include:

1. *recA*⁻ mutants. Mutations in the *recA* gene lead to loss of 98% of homologous recombination activity. This is an especially important mutation to have in the recipient strain when cloning homologous DNA, for example, when cloning *E. coli* genes, when maintaining two plasmids with regions of homology, or to prevent loss of direct repeats.

2. *polA⁻* mutants. The *polA* gene codes for DNA polymerase I, an enzyme required for the ColE1-type plasmid of replication. If you are using a vector with a ColE1-type origin of replication, it will not replicate in a *polA* mutant.

3. *dam⁻* mutants. These mutants are deficient in methylation of the adenine in the sequence 5'-GATC-3'. This methylase provides strand discrimination for mismatch repair that occurs after incorporation of the incorrect nucleotide during DNA replication. Several restriction enzymes recognize this sequence, and GATC is contained within or is a part of the recognition sequence of several others. Some of these enzymes will not cut if the sequence is methylated and others will cut only if the sequence is methylated. Therefore, DNA from *dam⁺* strains is resistant to some of these enzymes, and DNA from *dam⁻* strains is resistant to others.

4. *recB⁻ recC⁻ sbcB⁻* triple mutants. These mutants are useful for cloning palindromes or inverted repeats, which are commonly found in eukaryotic DNA (see Wyman and Wertman, 1987). Such a mutant is also used for cloning linear DNA in *E. coli* when replacing the wild-type gene on the chromosome with a mutation, insertion, or deletion of the gene by homologous recombination. A *recD* mutant is also useful for this purpose (Bick and Cohen, 1986). Mutations in the *recD* locus abolish the exonuclease activity of RecBCD (ExoV) without affecting recombination proficiency of the RecBCD pathway (Amundsen *et al.,* 1986).

B. Maintenance of bacterial stocks

To maintain pure cultures of bacterial strains, a single colony is picked with a sterile inoculating loop and streaked on an agar plate, diluting the bacteria in the process. Specifically, the loop is sterilized by flaming.

After the loop is cooled in the sterile agar, a single colony is touched with the loop, and the loop is drawn back and forth across an agar plate as shown in Fig. 1-2. The plate is turned 90°, the loop is sterilized and cooled, and the loop is drawn through the inoculated part of the plate once and then streaked back and forth on an uninoculated region. This process is repeated until most of the plate is covered. After growing the cultures at 37°C overnight, there should be isolated colonies in one re-gion of the plate. This plate can be stored at room temperature for one week if wrapped with Parafilm. Longer storage in the refrigerator is suc-cessful for only some strains of *E. coli*, but most strains can be stored semipermanently at room temperature in stock agar contained in airtight screw-capped tubes. Because of plasmid instability, strains containing plasmids should not be stored in this manner. Permanent storage at

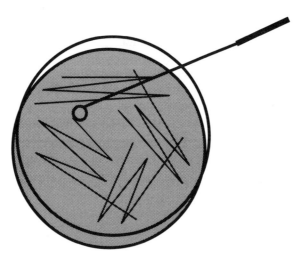

Figure 1-2. Method for diluting bacterial cells by streaking on an agar plate so that after incubation, single colonies will form rather than confluent growth.

–70°C is an essential and convenient way to store strains, especially those strains containing plasmids. Cells scraped from an overnight streak plate of the strain are mixed in M9 salts containing 40% glycerol in a screw-capped plastic vial made for storage at ultralow temperatures, and the vial is stored at –70°C. To obtain cells from the frozen stock culture, ice from the tube is scraped and streaked directly onto an agar plate. The tube is then returned immediately to the –70°C freezer.

C. Sources of strains

E. coli strains may be obtained from Dr. Barabara J. Bachmann, Curator, *E. coli* Genetic Stock Center, Department of Human Genetics, Yale University School of Medicine, 333 Cedar Street, New Haven, CT 06510. Some plasmid vectors may be obtained from the American Type Culture Collection, 12301 Parklawn Drive, Rockville, MD 20852. Vectors can sometimes be obtained commercially but are also generally available from the authors of papers describing the vectors. Some vectors were constructed in industrial settings and are only available commercially. See Appendix 4 for information about obtaining the strains and vectors used in this manual.

D. Use of sterile technique

Awareness of sterility is essential to the success of most of the experiments described in this manual. DNases and RNases are on your hands and are produced by bacteria that might be growing in unsterile buffers, so all materials used in the experiments must be sterilized by either autoclaving or filter sterilization. The antibiotic solutions must be sterilized by filtration because the antibiotics are heat labile. This is done by loading a syringe with the solution. A 0.2-μm sterile filter is attached to the syringe, then the solution is pushed through the filter into a sterile test tube. Of course, those organic compounds that are bactericidal, such as alcohol, phenol, and chloroform, are already sterile and should not be autoclaved.

For further discussion of maintenance and storage of bacteria and sterile technique, see Miller (1987). Because RNases are extremely resistant to heat, additional care must be taken to maintain an RNase-free environment. Refer to Blumberg (1987) if you are working with RNA.

REFERENCES

S. Amundsen, A. Taylor, A. M. Chaudhury, and G. R. Smith, *Proc. Natl. Acad. Sci. USA* **83,** 5558 (1986).

D. P. Bick and S. N. Cohen, *J. Bacteriol.* **167,** 594 (1986).

D. D. Blumberg, *Methods Enzymol.* **152,** 120 (1987).

J. L. Ingraham, O. Maaløe, and F. C. Neidhardt, *In* "Growth of the Bacterial Cell," Sinauer Associates, Sunderland, Massachusetts, 1983.

H. Miller, *Methods Enzymol.* **152,** 145 (1987).

X. Raleigh, *Methods Enzymol.* **152,** 130 (1987).

A. R. Wyman and K. F. Wertman, *Methods Enzymol.* **152,** 173 (1987).

ISOLATION AND ANALYSIS OF BACTERIAL AND *DROSOPHILA* CHROMOSOMAL DNA

There are several methods for isolating chromosomal DNA from cells. All have the common goal of obtaining high-molecular-weight material. Chromosomal DNA is quite long. For example, the chromosome of *E. coli* is 4.7×10^3 kb (Kohara, *et al.*, 1987), which is about 1 mm long. Because the double helix of DNA in the B form is only 10 Å wide, DNA is extremely susceptible to shearing forces. In any isolation procedure then, excessive agitation should be avoided. As a further precaution, ethylenediaminetetraacetic acid (EDTA) is included in the buffers to chelate Mg^{2+}. This will reduce deoxyribonuclease (DNase) activity because of the Mg^{2+} requirement of these enzymes. In the two procedures described here, the detergents sarcosyl and SDS (sodium dodecyl sulfate) are added to denature proteins and solubilize lipids in membranes leading to cell lysis. The cell lysate is often treated with enzymes that hydrolyze RNA and proteins. This is generally followed by phenol or phenol:chloroform extraction to remove the remaining proteins, which will either enter the organic phase or, if denatured, appear at the interphase. Ether treatment is used to remove the phenol, and alcohol precipitation concentrates the DNA while removing nucleotides, amino acids, and low-molecular-weight oligonucleotides and peptides.

Isolation and Purification of *E. coli* Chromosomal DNA

This procedure has been used successfully with *E. coli* and with several other types of Gram-negative bacteria, including *Rhizobia*, *Serratia*, and *Vibrio* sp. Chromosomal DNA isolation procedures for other microorganisms differ mainly in the lysis step, such as for *Bacillus* sp. (Saito and Miura, 1963) and for yeast (Holm *et al.*, 1986). The protocol described here results in DNA that is easily cut with restriction enzymes and, therefore, is useful for Southern hybridization studies. If the DNA is to be used to construct a clone bank, higher molecular weight DNA is desirable, so the vortex step should be eliminated.

MATERIALS

- ❐ *E. coli* strain LE392 *lacY galK2 galT22 metB1 trpR55 supE44 supF58 hsdR514* (r_k^-, m_k^+), grown in LB medium to late log phase (measure OD_{450} of culture in order to estimate the number of cells/ml)

- ❐ Container of ice

- ❐ TNE buffer (10 m*M* Tris–Cl, pH 8.0, 10 m*M* NaCl, 0.1 m*M* EDTA)

- ❐ TE buffer (10 m*M* Tris–Cl, pH 8.0, 0.1 m*M* EDTA)

- ❐ HTE buffer (50 m*M* Tris–Cl, pH 8.0, 20 m*M* EDTA)

- ❐ Sterile microcentrifuge tubes

- ❐ Sterile pipet tips

- ❐ Sterile Pasteur pipets

☐ RNase on ice (pancreatic RNase A, 10 mg/ml, in TE buffer, pre-heated to 80°C for 10 minutes to inactivate DNases)

☐ Pronase on ice (10 mg/ml in TNE buffer preheated to 37°C for 15 minutes to digest any DNases present)

☐ 2% Sarcosyl (*N*-lauroylsarcosine) in HTE buffer

☐ Phenol:chloroform (1:1, equilibrated with 0.5 *M* Tris–Cl, pH 8.0); the chloroform is a mixture of chloroform and isoamyl alcohol (24:1 vol/vol)

Caution: Phenol can cause severe skin burns. Wear gloves, a labcoat, and safety glasses when handling it. Work in a fume hood if possible. The laboratory should contain a solution of PEG 400 (diluted 1:1 with water); this is the solvent of preference for removing phenol from body surfaces (Pardoe *et al.*, 1976).

☐ 3.0 *M* sodium acetate

☐ Water bath set at 50°C

☐ Water bath set at 37°C

☐ Isopropanol (at –20°C)

☐ 70% ethanol (at room temperature)

☐ Ether (in hood)

☐ Polyethylene glycol (PEG) 400 diluted 1:1 with water—to be used for washing phenol burns

☐ Vortex mixer

☐ Dry ice–ethanol bath

☐ Vacuum dryer apparatus

PROTOCOL

1. Centrifuge 1.2 ml cells in microcentrifuge for 15 seconds.

2. Resuspend cell pellet in 0.31 ml HTE buffer making sure there are no cell clumps.

3. Add 0.35 ml 2% sarcosyl in HTE buffer. Mix well by inversion of tube.

4. Add 5 µl RNase, and incubate at 37°C for 15 minutes. Add 35 µl of pronase and heat at 50°C until lysis is complete (30–90 minutes).

5. Using the maximum speed setting, vortex the lysate for 2 minutes to shear the DNA. This step will lower the molecular weight of DNA molecules to around 20 kb, and, because of the decreased viscosity, it will be easier to extract the lysate with phenol:chloroform and ether. If you want to preserve the circularity of high-molecular-weight plasmids, do not vortex.

6. Phenol:chloroform extraction: Add an equal volume (0.70 ml) of phenol:chloroform 1:1, vortex briefly (few seconds) to mix, centrifuge at full speed in a microcentrifuge for three minutes to separate the phases, and carefully remove the *top* layer with a sterile Pasteur pipet and add to a new microcentrifuge tube. Be sure not to take along the white flocculent interphase. Re-extract the aqueous DNA layer with phenol:chloroform two more times. *Dispose of the phenol:chloroform in an approved waste receptacle.*

7. Ether extraction (*in hood*): Extract with an equal volume (0.7 ml) of ether— add ether, vortex briefly, and discard *top* layer in waste receptacle.

8. Precipitate DNA: Add sodium acetate to 0.3 M (70 µl), mix well, and add an equal volume of isopropanol (0.7 ml). After adding the isopropanol, invert tube five or six times. Freeze in an ethanol–dry ice bath for 10 minutes. Centrifuge for 5 minutes or longer (see note) and pour off supernatant.

9. Wash DNA pellet: Add ~1.0 ml of 70% ethanol, vortex briefly, centrifuge for 5 minutes, and pour off supernatant. Repeat two times. This removes residual salts as well as remaining phenol or

Note to 6: Phenol and chloroform will extract lipids and denature proteins, while isoamyl alcohol reduces foaming during the extraction and facilitates the separation of the aqueous and organic phases. Redistilled phenol should be stored at ⁻20°C to prevent the accumulation of contaminants that cause breakdown or crosslinking of RNA and DNA. If phenol has any color, it should be redistilled before use in nucleic acid purification. Phenol or phenol:chloroform solutions should be mixed with a buffer solution prior to use in extraction procedures. The lower (organic) phase is water-saturated and will, therefore, not absorb part or all of the aqueous phase containing the nucleic acids that you are purifying. See Wallace (1987) for discussion of choice of conditions for phenol extractions.

Caution in 7: Ether is used to extract traces of phenol and chloroform from the aqueous layer. Ether is highly volatile and extremely flammable; it should be used in a chemical hood.

Note to 8: If tube is placed in a fixed-angle rotor so that the top is oriented with the hinge to the outside, then the pellet is more easily located. If you have trouble with recoveries of DNA, try increasing the length of time you centrifuge. See Zeugin and Hartley (1985) for a critical examination of conditions affecting recovery of DNA after ethanol precipitation.

chloroform. Wrap top of tube with Parafilm, poke holes in Parafilm with needle, then dry pellet under vacuum to remove ethanol.

10. Dissolve the precipitated DNA in 50 μl of TE buffer. Label the tube and store it at –20°C.

Isolation and Purification of *Drosophila* DNA

This isolation procedure yields DNA of sufficient molecular weight for Southern blot analysis, so it is useful for mutant screening. If higher molecular weight DNA is required, for example, for constructing a genomic clone bank, follow the isolation procedure described by Herrmann and Frischauf (1987).

MATERIALS

☐ Adult *Drosophila* (either newly collected or stored frozen at –20°C)

☐ Container of ice

☐ TE buffer

☐ Ground glass homogenizers (2–10 ml capacity)

☐ Fly homogenization buffer (0.1 M Tris–Cl, pH 9.1, 0.1 M NaCl, 0.2 M sucrose, 0.05 M EDTA, 0.5% SDS)

☐ RNase on ice (pancreatic RNase A, 10 mg/ml, in TE buffer, pre-heated to 80°C for 10 minutes to inactivate DNases)

☐ 8 M potassium acetate

☐ Phenol (equilibrated with 0.5 M Tris–Cl, pH 8.0)

❐ Chloroform

❐ 95% ethanol (room temperature)

❐ 70% ethanol (room temperature)

❐ Water bath set at 65°C

PROTOCOL

1. Transfer 50 flies to a homogenizer and add 1 ml of homogenization buffer. Grind the flies for 20 seconds (until no whole flies remain). Vortex as necessary.

2. Divide the mixture into two 1.5-ml sterile microcentrifuge tubes.

3. Add 25 μl of RNase to each tube.

4. Hydrolyze the RNA and continue the tissue lysis by incubating the samples at 65°C for 30 minutes.

5. Precipitate the SDS and insoluble debris by adding 75 μl of 8 M potassium acetate to each tube. Mix immediately and put on ice for 30–60 minutes.

6. Centrifuge the samples at 4°C for 5 minutes in a microcentrifuge.

7. Transfer the supernatant from each tube into a clean microcentrifuge tube.

8. Remove protein contaminants from the samples by adding 400 μl of phenol to each tube and vortexing vigorously. Then add 400 μl of chloroform to each tube and vortex again.

9. Separate the aqueous and organic phases by centrifuging the samples at room temperature (in a microcentrifuge) for 5 minutes.

10. Carefully remove the aqueous (upper) layer from each tube and

transfer it to fresh tubes. **Do not** transfer the denatured white protein that appears at the interface between the two phases.

11. Re-extract the aqueous layer with 800 μl of chloroform (to remove phenol and additional protein). After vortexing and centrifugation, transfer the upper layers to fresh tubes.

12. Precipitate the DNA by adding one volume (~400 μl) of room temperature 95% ethanol.

13. Mix well and centrifuge in a microcentrifuge for 5 minutes.

14. Decant the ethanol and add ~1 ml of 70% ethanol to each tube. Vortex briefly and recentrifuge as in step 13. This removes residual salts as well as remaining phenol or chloroform.

15. Dry the pellets *in vacuo* and resuspend each pellet in 50 μl TE buffer. Store samples frozen at –20°C.

See Herrmann and Frischauf (1987) for procedures to isolate genomic DNA from mammalian cells, including whole organs, cell culture, or human blood.

Note: Dispose of the organic waste material in an approved container.

DAY 2

Determination of the Concentration and Purity of DNA by UV Spectroscopy

Both DNA and RNA absorb ultraviolet (UV) light so efficiently that absorbance spectroscopy can be used as an accurate, rapid, and non-destructive method to determine concentrations as low as 2.5 μg/ml. The nitrogenous bases in oligonucleotides have an absorption maximum at approximately 260 nm. Using a 1-cm light path, the extinction coefficient for DNA at this wavelength is 20. Based on this extinction coefficient, the absorbance at 260 nm in a 1-cm quartz cuvette of a 50 μg/ml solution of double-stranded DNA is equal to 1 (see Eq. 2-1).

Proteins absorb maximally at approximately 280 nm due mainly to tryptophan residues. The ratio of 260/280, therefore, is a measure of the purity of a DNA preparation and should fall between 1.65 and 1.85. A lower value suggests protein contamination. If phenol, which has one λ_{max} at 270 nm, is contaminating the DNA preparation, then the A_{260} will be abnormally high, leading to an overestimation of the DNA concentration.

MATERIALS

- ❐ Matching pairs of 1-ml quartz cuvettes
- ❐ TNE buffer
- ❐ Microcentrifuge tubes
- ❐ UV spectrophotometer (UV lamp prewarmed)

Note: Glass or plastic cuvettes absorb UV, so quartz cuvettes must be used to measure DNA and RNA concentrations.

PROTOCOL

1. Prepare a dilution of chromosomal DNA by adding 20 μl of your chromosomal DNA solution to 0.98 ml of TNE buffer in a microcentrifuge tube. Mix well.

2. Allow 20 minutes for the UV lamp in the spectrophotometer to warm up. Set the wavelength of the spectrophotometer to 260 nm. Add TNE buffer to one cuvette; use the TNE buffer as a blank and set the absorbance to 0. Measure the absorbance of your DNA dilution and write it down.

3. Repeat at 280 nm.

4. Calculate the concentration of DNA in your dilution, assuming that DNA at a concentration of 50 µg/ml has an OD of 1 at 260 nm. Calculate the DNA concentration in your original DNA solution. **Label** the tube containing your DNA with the date, contents, concentration, and your initials, then cover your writing with transparent tape for preservation. Store the DNA in a freezer box at –20°C.

$$\text{DNA concentration} = (OD_{260}) \times (\text{dilution factor})$$
$$\times (50 \text{ µg DNA}/1 \text{ } OD_{260} \text{ unit}) \quad\quad (2\text{-}1)$$

5. Calculate total yield:

$$\# \text{ µg DNA recovered} = (\text{DNA concentration in µg/ml})$$
$$\times (\text{total volume in ml}) \quad\quad (2\text{-}2)$$

6. Calculate the expected yield of DNA. Use the following values for *E. coli*: MW of the *E. coli* chromosome is 3.1×10^9; Avogadro's number is 6.02×10^{23} molecules/mole; the culture volume is 1.2 ml; in late log phase, there are approximately 2 chromosomes/cell.

E. coli, expected yield:

$$\# \text{ chromosomes} = \text{vol (ml)} \times (\text{cells/ml})$$
$$\times (\# \text{ chromosomes/cell}) \quad\quad (2\text{-}3)$$

$$\text{Expected yield} = \frac{3.1 \times 10^9 \text{ g/mol}}{6.02 \times 10^{23} \text{ chromosomes/mol}} \times \# \text{ chromosomes} \quad (2\text{-}4)$$

For *Drosophila*, calculation of the expected DNA yield is difficult because various fly tissues have different chromosome ploidy, e.g., some are haploid, others are diploid, and still others are polyploid. Practical experience indicates that each fly yields approximately 0.3 µg of DNA, and this value should be utilized for calculating the expected yield of DNA.

7. Calculate the % recovery of DNA for both *E. coli* and *Drosophila*:

$$\% \text{ recovery} = (\text{actual yield/expected yield}) \times 100 \quad\quad (2\text{-}5)$$

8. Calculate the number of haploid genomes of *Drosophila* DNA you have obtained (ignoring the contribution of mitochondrial DNA). The number of daltons per haploid *Drosophila* genome is 1.1×10^{11}.

$$\text{Number of haploid genomes} = \text{g of DNA recovered} \div \text{g/haploid genome} \qquad (2\text{-}6)$$

$$\text{g/haploid genome} = \frac{1.1 \times 10^{11} \text{ g/mol}}{6.02 \times 10^{23} \text{ molecules/mol}} \qquad (2\text{-}7)$$

9. Estimate the purity of your DNA preparation from the OD_{260}/OD_{280} ratio.

Restriction Endonuclease Digestion of Chromosomal DNA

Most type II restriction endonucleases recognize and cut DNA within or near specific sequences. See Table 2-1 for sequence specificity of commercially available restriction endonucleases. One such enzyme, *Bam*HI, will be used in this exercise to cut the *E. coli* or *Drosophila* chromosomal DNA that was isolated previously. *Bam*HI recognizes the following sequence with the arrows indicating the exact sites of cleavage:

```
5'-G G A T C↓C-3'
3'-C↑C T A G G-5'
```

This hexameric sequence has a probability of appearing $1/4^6$ bp or once every 4096 bp. Hence, the average size of a piece of DNA or restriction fragment generated by *Bam*HI is around 4 kb. See Brooks (1987) for a detailed discussion of the properties and uses of restriction

Table 2-1. Restriction Endonucleases[a]

Enzyme	Recognition Sequence	Enzyme	Recognition Sequence	Enzyme	Recognition Sequence
AatI	AGGCCT	AvrII	C↓CTAGG	DdeI	C↓TNAG
AatII	GACGT↓C	BalI	TGG↓CCA	DraI	TTT↓AAA
AccI	GT↓MKAC	BamHI	G↓GATCC	DraII	RG↓GNCCY
AccII	CG↓CG	BanI	G↓GYRCC	DraIII	CACNNN↓GTG
AccIII	T↓CCGGA	BanII	GRGCY↓C	DpnI	GmA↓TC
AcyI	GR↓CGYC	BanIII	ATCGAT	EaeI	Y↓GGCCR
AflI	G↓GWCC	BbeI	GGCGC↓C	"	C↓GGCCG
AflII	C↓TTAAG	BbiII	GR↓CGYC	EagI	C↓GGCCG
AflIII	A↓CRYGT	BbvI	GCAGC(8/12)	Eco47I	G↓GWCC
AhaII	GR↓CGYC	BbvI	GCTGC-(12/8)	Eco47III	AGC↓GCT
AhaIII	TTT↓AAA	BclI	T↓GATCA	Eco52I	C↓GGCCG
AluI	AG↓CT	BcnI	CC↓SGG	Eco8II	CC↓TNAGG
AlwI	GATCC-(5/4)	BglI	GCCN₄↓NGGC	EcoO109I	RG↓GNCCY
"	GGATC(4/5)	BglII	A↓GATCT	EcoRI	G↓AATTC
AlwNI	CAGN₃↓CTG	BsmI	GAATGC(1/-1)	EcoRII	↓CCWGG
AocI	CC↓TNAGG	BsmI	GCATTC-(-1/1)	EcoRV	GAT↓ATC
AocII	GDGCH↓C	Bsp1286I	GDGCH↓C	EcoT22I	ATGCA↓T
AosI	TGC↓GCA	BspHI	T↓CATGA	EcoT14I	C↓CWWGG
AosII	GR↓CGYC	BspMI	ACCTGC	EcoVIII	A↓AGCTT
ApaI	GGGCC↓C	BspMII	T↓CCGGA	EspI	GC↓TNAGC
ApaLI	G↓TGCAC	BssHII	G↓CGCGC	Fnu4HI	GC↓NGC
ApyI	CC↓WGG	BstEII	G↓GTNACC	FnuDII	CG↓CG
AquI	C↓YCGRG	BstI	G↓GATCC	FokI	CATCC-(13/9)
Asp700I	GAANN↓NNTTC	BstNI	CC↓WGG	FspII	TT↓CGAA
Asp718I	G↓GTACC	BstUI	CG↓CG	FspI	TGC↓GCA
AsuI	G↓GNCC	BstXI	CCAN₅↓NTGG	HaeII	RGCGC↓Y
AsuII	TT↓CGAA	CfoI	GCGC	HaeIII	GG↓CC
AvaI	C↓YCGRG	Cfr10I	R↓CCGGY	HapII	C↓CGG
AvaII	G↓GWCC	Cfr13I	G↓GNCC	HgaI	GACGC(5/10)
AvaIII	ATGCAT	ClaI	AT↓CGAT	"	GCGTC-(10/5)
AvrI	CYCGRG	CvnI	CC↓TNAGG	HgiAI	GWGCW↓C

[a]The enzymes listed above are commercially available. The following conventions are used. Directions and abbreviations: All sequences are written 5' to 3'. Incompletely specified bases are abbreviated as given: R=A or G, K=G or T, H=A or C or T, D=A or G or T, Y=C or T, S=G or C, B=C or G or T, N=A or C or G or T, M=A or C, W=A or T, V=A or C or G. Cleavage position: ↓ indicates the point of cleavage. If ↓ does not appear with the recognition sequence, then the cleavage point is either outside the sequence, or is unknown. Numbers in parentheses indicate the number of bases between the end of the recognition sequence and the cleavage point. The first number is for the strand shown, the second number is for the opposite strand. Minus sign indicates that cleavage occurs to the left of the recognition site. Lower case m indicates following base is methlyated. Adapted from the BRL catalog. *(Table continues.)*

21

Table 2-1. Restriction Endonucleases *(continued)*

Enzyme	Recognition Sequence	Enzyme	Recognition Sequence	Enzyme	Recognition Sequence
*Hha*I	GCG↓C	*Not*I	GC↓GGCCGC	*Sca*I	AGT↓ACT
*Hinc*II	GTY↓RAC	*Nru*I	TCG↓CGA	*Scr*FI	CC↓NGG
*Hind*II	GTY↓RAC	*Nsi*I	ATGCA↓T	*Sdu*I	GDGCH↓C
*Hind*III	A↓AGCTT	*Nsp*(7524)I	RCATG↓Y	*Sec*I	C↓CNNGG
*Hinf*I	G↓ANTC	*Nsp*BII	CMG↓CKG	*Sex*I	CTCGAG
*Hin*P1I	G↓CGC	*Nsp*HI	RCATG↓Y	*Sfa*NI	GCATC(5/9)
*Hpa*I	GTT↓AAC	*Nsp*II	GDGCH↓C	"	GATGC−(9/5)
*Hpa*II	C↓CGG	*Nsp*III	C↓YCGRG	*Sfi*I	GGCCN$_4$↓NGGCC
*Hph*I	GGTGA(8/7)	*Nsp*IV	G↓GNCC	*Sin*I	G↓GWCC
*Kpn*I	GGTAC↓C	*Nsp*V	TTCGAA	*Sma*I	CCC↓GGG
*Mae*I	C↓TAG	*Nun*II	GG↓CGCC	*Sna*BI	TAC↓GTA
*Mae*II	A↓CGT	*Oxa*NI	CC↓TNAGG	*Spe*I	A↓CTAGT
*Mae*III	↓GTNAC	*Pae*R7I	C↓TCGAG	*Sph*I	GCATG↓C
*Mbo*I	↓GATC	*Pal*I	GG↓CC	*Spl*I	C↓GTACG
*Mbo*II	GAAGA(8/7)	*Pfl*MI	CCAN$_4$↓NTGG	*Ssp*I	AAT↓ATT
*Mfl*I	R↓GATCY	*Ple*I	GACTC−(5/4)	*Sst*I	GAGCT↓C
*Mlu*I	A↓CGCGT	"	GAGTC (4/5)	*Sst*II	CCGC↓GG
*Mnl*I	CCTC(7/7)	*Ppu*MI	RG↓GWCCY	*Stu*I	AGG↓CCT
"	GAGG−(7/7)	*Pss*I	RGGNC↓CY	*Sty*I	C↓CWWGG
*Msp*I	C↓CGG	*Pst*I	CTGCA↓G	*Taq*I	T↓CGA
*Mst*II	CC↓TNAGG	*Pvu*I	CGAT↓CG	*Tha*I	CG↓CG
*Mva*I	CC↓WGG	*Pvu*II	CAG↓CTG	*Tth*HB8I	T↓CGA
*Nae*I	GCC↓GGC	*Rsa*I	GT↓AC	*Tth*lllI	GACN↓NNGTC
*Nar*I	GG↓CGCC	*Rsr*I	G↓AATTC	*Xba*I	T↓CTAGA
*Nci*I	CC↓SGG	*Rsr*II	CG↓GWCCG	*Xcy*I	C↓CCGGG
*Nco*I	C↓CATGG	*Sac*I	GAGCT↓C	*Xho*I	C↓TCGAG
*Nde*I	CA↓TATG	*Sac*II	CCGC↓GG	*Xho*II	R↓GATCY
*Nde*II	↓GATC	*Sal*I	G↓TCGAC	*Xma*I	C↓CCGGG
*Nhe*I	G↓CTAGC	*Sau*3AI	↓GATC	*Xma*III	C↓GGCCG
*Nla*III	CATG↓	*Sau*96I	G↓GNCC	*Xmn*I	GAANN↓NNTTC
*Nla*IV	GGN↓NCC	*Sau*I	CC↓TNAGG	*Xor*II	CGAT↓CG

endonucleases. Type II restriction enzymes, which can be purchased commercially, are expensive and labile. Because they are *very* easily destroyed, the following rules should help in preserving them:

1. Write out the protocol to be followed for the enzyme digestion, including the amount of each component to be added to the reaction mixture, DNA concentration, and time and temperature of incubation, and check off each component after its addition.

2. Wash hands before using an enzyme, and, if available, wear a fresh pair of gloves when handling the enzyme tube. Do not touch the inner surfaces of the tube, for example, the inside surface of the lid, because skin contains proteases and nucleases. If available, use a tube opener to open the tube containing the enzyme.

3. Always add the enzyme as the last component of the reaction mixture.

4. Do not remove the enzyme from the freezer until you are ready to add it to the reaction mixture. Return the enzyme to the freezer as soon as you are finished adding it to the reaction mixture.

5. When an enzyme is removed from the freezer, carry it to your bench in an ice bucket. Never allow an enzyme to be exposed to room temperature.

6. Always use fresh *sterile* pipet tips to add enzyme to your reaction mixture. If you use a dirty tip, you will have contaminated the enzyme for the next users.

MATERIALS

☐ 10x *Bam*HI buffer (500 mM Tris–Cl, pH 8.0, 100 mM MgCl$_2$, 1000 mM NaCl) (check manufacturer's conditions)

☐ *Bam*HI (~5 units/µl), stored in microcentrifuge tube at -20°C.

☐   Sterile distilled water

☐   37°C water bath

☐   Ice bucket

☐   Chromosomal DNA isolated from *E. coli* or *Drosophila*

PROTOCOL

1. Digestion of the chromosomal DNA isolated previously: Use the volume necessary to add 1–3 µg of your chromosomal DNA to the reaction mix. Calculate the amount to add from the OD_{260} measurements you made. Be sure to leave at least 1 µl of DNA in the tube to be run on the agarose gel as the undigested control for the next protocol.

2. *Add in the following order* the components of the restriction digest reaction mixture to a microfuge tube (final volume of 50 µl):

   a. _____ µl of sterile distilled water

   b. 5   µl of 10x *Bam*HI buffer

   c. _____ µl of chromosomal DNA (1–3 µg)

   d. 1   µl of *Bam*HI restriction endonuclease (5 units)

   MIX WELL since restriction enzymes are provided in glycerol-containing buffers and glycerol is more dense than water.

3. Allow the reaction mixture to digest for 1–2 hours or overnight at 37°C. Store the reaction in a –20°C freezer until ready to proceed.

Agarose Gel of Chromosomal DNA Restriction Endonuclease Digestions

DNA is negatively charged and will migrate to the positive electrode (anode) in an electric field. Because DNA molecules have a uniform charge:mass ratio, they have the same electrophoretic migration properties in a resistance-free medium. Agarose and acrylamide gels provide viscosity such that DNA molecules will separate according to size and conformation during electrophoresis. The linear DNA molecules generated in the *Bam*HI digestion of chromosomal DNA will be separated according to size in an 0.8% agarose gel in this experiment. DNA is visualized using the intercalating dye, ethidium bromide, which fluoresces when irradiated with UV light. The size of the fragments generated can be estimated by comparing the electrophoretic mobility (distance migrated through the agarose gel per unit time) of an unknown DNA molecule to the electrophoretic mobilities of DNA molecules for which the sizes are known. For DNA molecules of the same conformation, there is a linear relationship between the logarithm of their molecular weights expressed either in kDa or kb and their electrophoretic mobilities, especially for sizes smaller than 8 kb. Generally, a standard curve is drawn as described below on semilog paper from which the size of an unknown fragment is estimated. See Ogden and Adams (1987) for details concerning choice of gel systems, concentration of the solid matrix and electrophoretic conditions to be used for separating different-sized DNA molecules, convenient size markers, techniques for eluting DNA from agarose gels, and gel drying conditions.

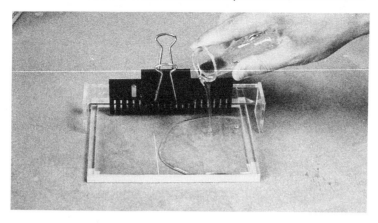

Figure 2-1. Photo of comb placement and gel pouring.

MATERIALS

❏ Loading buffer (0.25% bromphenol blue, 0.25% xylene cyanol, and 30% glycerol)

❏ 10x TEA (per liter: 48.4 g of Trisma base, 11.42 ml of glacial acetic acid, 20 ml of 0.5 M EDTA, pH 8.0)

❏ 0.8% agarose in flasks in 60°C water bath (100 ml/flask of: 0.8 g of agarose, 10 ml of 10x TEA, 90 ml water; boil for a few minutes to dissolve agarose, then store at 60° C until ready to pour gel)

❏ Gel frames (10 cm long) and well-formers (combs)

❏ DNA size markers [1-kb DNA ladder, obtained from Bethesda Research Laboratories (BRL)]

PROTOCOL

1. Seal the outside of the gel former with tape. Position the comb about 1 mm above the surface of the frame by placing an index card

Caution: It is important that the comb not touch the bottom of the gel frame while the gel is being poured; otherwise the sample well will have no agarose bottom and the sample will leak underneath the gel.

between the bottom of the comb and the gel plate. Remove the index card and pour 0.8% agarose into the gel frame. After the gel has hardened completely (20–30 minutes), gently remove the comb. Transfer the gel to the electrophoretic apparatus and cover it with 1x TEA buffer.

2. In this protocol, you will run two DNA samples: undigested chromosomal DNA and chromosomal DNA digested with *Bam*HI. (You may, of course, run as many samples as you have wells.) The DNA samples are mixed with a droplet of loading buffer on Parafilm, which has a hydrophobic surface. After removing the paper covering the Parafilm, use the clean Parafilm surface facing the paper to mix samples for loading.

   a. Undigested sample: mix 2 µl of DNA with 2 µl of loading buffer on Parafilm.

   b. *Bam*HI digested sample: mix 25 µl of DNA with 5 µl of loading buffer on Parafilm.

3. Load 0.5 µg of the 1-kb ladder in one well. Using a micropipettor, carefully load each sample into a well. The density of the glycerol solution will cause the sample to sink to the bottom of the well. *Record what is in each well.*

   Caution: Be certain that the safety lid is in place prior to turning on the power supply. Never touch a gel that has current passing through it.

4. Connect the electrophoresis apparatus to the power supply and turn on the power supply. Always record the agarose concentration, voltage used, distance migrated of the two dyes, and time of electrophoresis for each gel you run; this will aid in determining conditions for separation of similar DNA digests in the future. Approximate electrophoresis times for a 12-cm submerged 0.8% agarose gel are 3 hours at 120 V or overnight at 20 V.

5. When the deep blue dye (bromphenol blue) has migrated to the bottom of the gel, proceed to staining and photography.

DAY 5

Staining and Photography of Agarose Gel of Chromosomal DNA Restriction Endonuclease Digestions

MATERIALS

☐ Ethidium bromide solution made up fresh in water (1 µg/ml); prepare using a stock of 10 mg/ml (stored in light-protected container) diluted 1:10,000

Caution: Ethidium bromide is a mutagen and should not be allowed to come in contact with skin. Wear gloves when using ethidium bromide.

☐ UV (302- or 356-nm wavelength) transilluminator box

☐ Polaroid MP4 camera on stand or with shade; use Kodak Wratten red 23A and orange filters

☐ Type 57 high-speed print film

☐ Plastic wrap

☐ Fluorescent rulers (can be obtained from Diversified Biotech, 46 Marcellus Drive, Newton Centre, MA 02159). Clear plastic rulers may be used instead, but it is more difficult to see the ruler markings.

PROTOCOL

1. Stain the gel for 20 minutes in ethidium bromide solution.

2. Destain the gel in water for 20 minutes.

3. Wet the UV transilluminator box with water, then cover it with plastic wrap. The plastic wrap prevents ethidium bromide staining of the UV box. Carefully move the gel to the UV box. Place clear

Note: An alternative to staining the gel after electrophoresis is to add the ethidium bromide at a final concentration of 0.5 µg/ml to the anode (+) buffer reservoir. Ethidium bromide, which is positively charged, will stain the DNA as it moves through the agarose gel to the cathode (−). DNA, which is negatively charged, moves in the opposite direction.

plastic rulers on each side of the gel so that the distance each DNA band has moved from the top of the well can be estimated.

4. After putting on UV protective glasses, turn on the UV light and examine the gel. Take a photograph with the Polaroid camera.

5. Use the resulting photograph to determine the distance migrated for each band in the 1-kb ladder and prepare a semilog plot of the re-

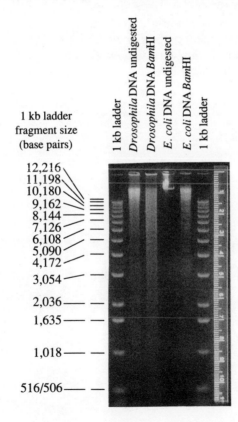

Figure 2-2. Photo of agarose gel of chromosomal DNA.

striction fragment size (plot kb on the log scale) against the distance migrated (on the linear scale). Use this plot to estimate the range of sizes seen in your *Bam*HI digestion of chromosomal DNA. Does this range correspond to the expected average size of a *Bam*HI restriction fragment? Estimate the expected number of unique *Bam*HI fragments in the *E. coli* chromosome (4.7×10^3 kb) and in the four *Drosophila* chromosomes, assuming that the sequence cut by *Bam*HI occurs once every 4096 bp. If the observed number of *Bam*HI sites in the *E. coli* chromosome is 472 (Kohara *et al.*, 1987), the average size of a *Bam*HI fragment in the *E. coli* chromosome should be 10 kb rather than 4 kb. Is this what you observe in your *Bam*HI digestion of *E. coli* chromosomal DNA?

REFERENCES

J. E. Brooks, *Methods Enzymol.* **152**, 113 (1987).

B. G. Herrmann and A.-M. Frischauf, *Methods Enzymol.* **152**, 180 (1987).

C. Holm, D. W. Meeks-Wagnor, W. L. Fangman, and D. Botstein, *Gene* **42**, 169 (1986).

Y. Kohara, K. Akiyama, and K. Isono, *Cell* **50**, 495 (1987).

R. Pardoe, R. T. Minami, R. M. Sato, and S. L. Schlesinger, *Burns* **3**, 29 (1976).

R. C. Ogden and D. A. Adams, *Methods Enzymol.* **152**, 61 (1987).

H. Saito and K. Miura, *Biochim. Biophys. Acta* **72**, 619 (1963).

D. M. Wallace, *Methods Enzymol.* **152,** 33 (1987).

J. A. Zeugin and J. H. Hartley, *Focus* **7**(4), 1 (1985).

PLASMID DNA ISOLATION AND AGAROSE GEL ANALYSIS

The first vectors constructed for cloning of recombinant DNA molecules were plasmids that contained the pMB1 origin of replication, for example, pBR322 (Bolivar *et al.*, 1977; Sutcliff, 1979; Balbas *et al.*, 1986) and pUC19 (Yanisch-Perron *et al.*, 1985). The pMB1 origin of replication is almost identical in sequence to the ColE1 origin of replication. These two plasmids are incompatible with each other in that they cannot be stably maintained in the same cell. Plasmid **incompatibility** is linked to the process by which the plasmids control their rate of self-duplication. Because incompatibility indicates a degree of similarity between plasmids, plasmids are divided into incompatibility groups.

Host range is another property of plasmids. Plasmids can have a narrow host range, replicating in very few closely related bacteria, or they can have a broad host range, replicating in widely diverse bacteria. Once again, the origin of replication is responsible for this property. Those vectors in the ColE1 incompatibility group have a narrow host range, replicating only in *E. coli* or closely related bacteria.

A third property of plasmids is **copy number**. Plasmids are maintained at a high, medium, or low number of copies per cell. When DNA yields are important, the high copy number vectors related to ColE1 are most useful. Popular cloning vectors of this type include the pUC plasmids, 7, 8, 9, 12, 13, 18, and 19 (Vieira and Messing, 1982; Norrander *et al.*,

1983; Yanisch-Perron *et al.*, 1985), the Bluescript plasmids and phage-mids of Stratagene and the pGEM® plasmid and phagemid vectors of Promega. The phagemids of Stratagene and Promega are derived from pUC19 and contain the origin of replication from f1, a single-stranded bacteriophage. They also contain the SP6, T3, or T7 promoters flanking the multiple cloning site (MCS, see Chapter 6, Fig. 6-2) that make possible the synthesis of RNA complementary to either the coding or non-coding strand of DNA inserted into the MCS.

When cloned DNA encodes peptides that cause death or slow growth in *E. coli*, low copy number vectors are required. Some of the most useful vectors of the latter kind are derived from the broad host range plasmid, R1. At 30°C, these plasmid vectors are maintained at one copy per cell; at 42°C, runaway replication occurs, leading to dramatic amplification of the copy number (Uhlin *et al.*, 1979; Larsen *et al.*, 1984).

An advantage of using high copy number plasmids related to ColE1 is that replication of these plasmids does not depend on proteins encoded by the plasmid; consequently, these plasmids will replicate in the presence of protein synthesis inhibitors such as chloramphenicol or spectino-mycin. The initiation process of *E. coli* chromosomal replication requires protein synthesis, so upon inhibition of protein synthesis, plasmid replication is amplified, resulting in an increase of from 30 to 50 copies per cell to 300–2000 copies per cell. This is especially useful when isolating large amounts of plasmid DNA as described in the appendix.

In this exercise, you will isolate plasmid DNA from a small number of cells; the DNA isolated in this manner is often referred to as a **miniprep**. The plasmid to be isolated, pBR329 (Covarrubias and Bolivar, 1982; Fig. 3-1), contains genes encoding resistance to three antibiotics: ampicillin (Ap^r), chloramphenicol (Cm^r), and tetracycline (Tc^r). Several restriction enzyme cloning sites appear only once in pBR329 (Fig. 3-1,

Table 3-1). The advantage of using restriction enzymes that cut only once in a cloning vector such as pBR329 is that the process of religation with the DNA to be inserted does not involve vector reconstruction. If the site is within an antibiotic resistance gene, then insertion of a DNA fragment into the site will inactivate the gene, thereby producing an antibiotic-sensitive phenotype. For example, if you choose to insert DNA cut with *Bam*HI into the *Bam*HI site of pBR329, which is in the *tet* gene, then the resulting plasmid will be Ap^r, Cm^r, but Tc^s. Transformants can be selected on agar plates containing ampicillin and chloramphenicol, then after colonies have developed, they can be screened for tetracycline sensitivity. Those that are Tc^s will contain DNA inserted into the *tet* gene. The nucleotide sequence of pBR329 is known (see Covarrubias and Bolivar, 1982), so new restriction enzyme sites can be identified using a computer program designed for such a purpose, as described in the chapter on the use of computers.

The most useful high copy number cloning vectors contain at least one gene for selection of the plasmid and a method to screen for DNA inserts into the plasmid. If the plasmid contains more than one antibiotic resistance marker, then one can be used for selection and the other for insertional inactivation. A wide variety of such plasmids is available, including pBR322 Ap^r, Tc^r (Balbas *et al.*, 1986) and pMK2004 Ap^r, Km^r, Tc^r (Kahn *et al.*, 1979). Two very useful plasmids, pACYC177 Ap^r, Km^r and pACYC184 Cm^r, Tc^r (Chang and Cohen, 1978), have similar properties to pBR322, pBR329, and pMK2004, except that they are compatible with these pMB1-derived plasmids. This means, for example, that pBR322 and pACYC184 can stably coexist in the same cell when selecting for resistance to ampicillin and chloramphenicol. Two compatible plasmids, each containing different cloned genes, have been used to examine the effects of one cloned gene product on the expression of the

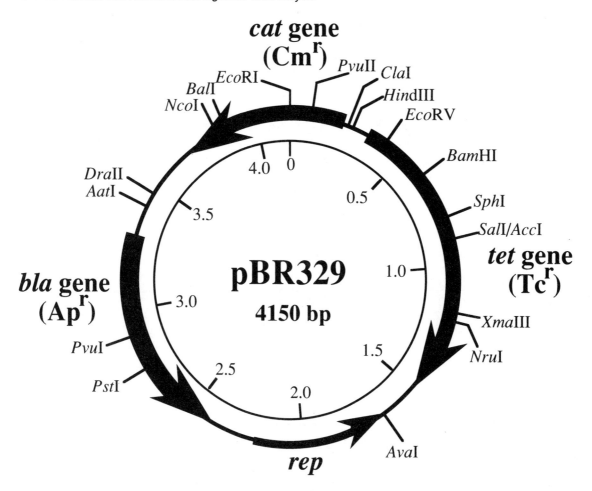

Figure 3-1. Genetic and restriction enzyme map of plasmid pBR329. Arrows cover coding region of genes except for *rep*, where the arrow denotes the region required for DNA replication and the direction of replication.

other cloned gene product or to selectively express one gene product whose expression is controlled by the other cloned gene product (see example in Tabor and Richardson, 1985).

In the following experiment, the strain from which pBR329 will be isolated is HB101::Tn5. The transposon, Tn5, is inserted at an unknown location in the host chromosome. Transposition of Tn5 to a new location occurs at a very low frequency (10^{-5}), so a small number of the pBR329 molecules isolated in this procedure will contain Tn5. The number is sufficiently low that the molecules containing Tn5 will not be seen on the agarose gel to be run on Day 2. The miniprep of pBR329 will be used in Lab IV to transform *E. coli*.

Table 3-1. Enzymes That Cut Once in pBR329

Enzyme	Site	Position
*Aat*II	G A C G T↓C	3427
*Acc*I	G T↓C G A C[a]	881
*Asu*II	T T↓C G A A	3619
*Ava*I	C↓T C G G G[a]	1655
*Bal*I	T G G↓C C A	3883
*Bam*HI	G↓G A T C C	606
*Cla*I	A T↓C G A T	254
*Dra*II	A G↓G C C C T[a]	3484
*Eco*RI	G↓A A T T C	4148
*Eco*RV	G A T↓A T C	416
*Hind*III	A↓A G C T T	260
*Nco*I	C↓C A T G G	3847
*Nru*I	T C G↓C G A	1202
*Pst*I	C T G C A↓G	2750
*Pvu*I	C G A T↓C G	2876
*Pvu*II	C A G↓C T G	100
*Sal*I	G↓T C G A C	881
*Sph*I	G C A T G↓C	792
*Xma*III	C↓G G C C G	1169

[a]This site is a subset of the sequences recognized by the restriction enzyme.

Isolation of Plasmid DNA by the Alkaline-Detergent Method—A Miniprep Procedure

MATERIALS

☐ LB agar plate streaked with strain HB101::Tn5 containing pBR329 and incubated at 37°C overnight

☐ Sterile microcentrifuge tubes

☐ Ice bucket with ice

☐ Solution I (25 mM Tris–Cl, pH 8.0, 50 mM glucose, 10 mM EDTA)

☐ Solution I containing 4 mg/ml lysozyme, freshly prepared (store on ice but warm to room temperature when ready to use)

☐ Solution II (0.2 N NaOH, 1.0% SDS, made fresh just prior to use)

☐ Solution III (3 M sodium acetate, 3 M acetic acid) stored on ice

☐ TE buffer containing 50 µg/ml RNase A (preheated to 80°C for 10 minutes to inactivate DNases)

☐ Isopropanol (room temperature)

☐ 70% ethanol

☐ Pasteur pipets and pipetting bulbs

☐ Phenol:chloroform (1:1) saturated with TE buffer

❏ Vortex mixer

❏ 37°C water bath

❏ Microcentrifuge

❏ Vacuum drying apparatus

PROTOCOL

1. Scrape a large loopful of cells from the agar plate of pBR329/HB101::Tn5 and resuspend in 200 μl of Solution I. Vortex vigorously.

2. Centrifuge the cells for 15–30 seconds in the microcentrifuge. Excess centrifugation will make the cells difficult to resuspend.

3. Remove the supernatant and resuspend the pellet in 200 μl of Solution I containing 4 mg/ml lysozyme (at room temperature). Vortex cells into suspension, making sure there are no clumps of cells.

4. After a few seconds at room temperature, lysozyme action on the cells will be visible: the suspended cells turn slightly milky and tend to coat the sides of the tubes when shaken. Place the tubes into an ice bucket and chill 1 minute.

5. Add 400 μl of Solution II at room temperature and mix by shaking the tube briefly. Return tube to the ice bucket for another minute.

6. Add 300 μl of ice-cold Solution III, mix, and return to the ice bucket for another 5 minutes. A fluffy white precipitate will form.

7. Centrifuge for 5 minutes in a microcentrifuge.

8. Pipett the supernatant into a fresh 1.5 ml microcentrifuge tube containing 0.6 ml of phenol:chloroform (1:1). Have these tubes

ready when the centrifuge stops. Absolutely avoid transferring any of the pellet, and try to avoid transferring any of the floating debris.

9. Vortex vigorously several seconds, then centrifuge 2 minutes at room temperature.

10. Pipet the aqueous upper layer into a fresh 1.5-ml microcentrifuge tube prefilled with 0.7 ml isopropanol (room temperature). Vortex briefly, then set at room temperature for 2 minutes.

11. Centrifuge at room temperature for 5 minutes. Pour off the supernatant. In order to remove as much liquid as possible, centrifuge the tube for 1 second, remove the rest of the supernatant without disturbing the pellet, and discard.

12. Wash pellet by gently mixing with 1 ml 70% ethanol and centrifuge for 3 minutes at room temperature.

13. Pour off the supernatant and dry the pellet under vacuum.

14. Resuspend the pellet in 50 µl of TE buffer containing 50 µg/ml of RNase A. Incubate at 37°C for 5 minutes. (In the absence of RNase treatment, DNA fragments smaller than about 600 bp are obscured by RNA when separated by agarose gel electrophoresis.)

15. Label the tube containing the miniprep of pBR329/HB101::Tn5 with date and contents, and store at −20°C.

Agarose Gel Electrophoresis of Undigested Plasmid DNA

Agarose gel electrophoresis can be used to resolve the different molecular configurations of a DNA molecule as well as to separate DNA fragments of different lengths. The gel you will use in this exercise will separate the supercoiled covalently closed circular DNA (Form I), the nicked or relaxed circular DNA (Form II), and the linear DNA (Form III) molecules in your plasmid DNA preparation. Under the conditions used (Tris acetate buffer), the compact supercoils will migrate faster than the linear DNA, which will migrate faster than the open circular Form II DNA. For a particular form of DNA, the migration rate in the agarose gel is inversely proportional to the logarithm of the molecular weight of the DNA molecule for a certain range of molecular weights. As a control, the supercoiled ladder from BRL containing plasmids of known size will be included on the agarose gel. You can use the distance migrated of the control plasmids to calculate a molecular weight standard curve from which to confirm the molecular weight of pBR329 supercoiled plasmid DNA that you isolated.

MATERIALS

- ❒ Loading buffer
- ❒ Horizontal 0.8% agarose gel
- ❒ Plasmid supercoiled ladder from BRL, diluted in loading buffer to 0.1 μg/μl
- ❒ Fluorescent rulers

PROTOCOL

1. Follow the procedure used for pouring gels previously described in Lab II, Day 4.

2. Mix 3 μl of pBR329/HB101::Tn5 with 3 μl of loading buffer cη a piece of Parafilm and add to one of the wells in the agarose gel.

3. Load 3 μl (0.3 μg) of supercoiled ladder in another well.

4. After running either overnight at 20 V or for ~2 hours at 200 V, stain the gel with 1 μg/ml ethidium bromide for 20 minutes, then destain the gel in water for 20 minutes.

5. Photograph the gel, and use the photograph to determine distance migrated for the supercoiled bands of the plasmid standards. Prepare a semilog plot of the plasmid size (plot kb on the log scale) against the distance migrated (on the linear scale). Use this plot to calculate the size of your plasmid.

6. Estimate the concentration of the DNA in your sample by comparing the intensity of pBR329 bands in the gel to the intensities of the bands in the supercoiled ladder, where a known amount of DNA was added.

Note: It is preferable to stain supercoiled DNA *after* agarose gel electrophoresis is complete rather than during electrophoresis, because ethidium bromide introduces positive supercoils into the negatively supercoiled plasmid molecules. The molecules will migrate differently in the presence of ethidium bromide than in its absence, depending on the concentration of ethidium bromide and the degree of initial negative superhelicity.

Note: For low amounts of DNA, less sample is wasted using an agarose gel to estimate the DNA concentration than when measuring the concentration in a spectrophotometer at 260 nm. For more accurate determinations, run several dilutions of a DNA sample for which the concentration is known.

REFERENCES

P. Balbas, X. Soberon, E. Merino, M. Zurita, H. Lomeli, F. Valle, N. Flores, and F. Bolivar, *Gene* **50**, 3 (1986).

F. Bolivar, R. L. Rodriguez, P. J. Greene, M. V. Betlach, H. L. Hdynecker, H. W. Boyer, J. H. Crosa, and S. Falkow, *Gene* **2**, 95 (1977).

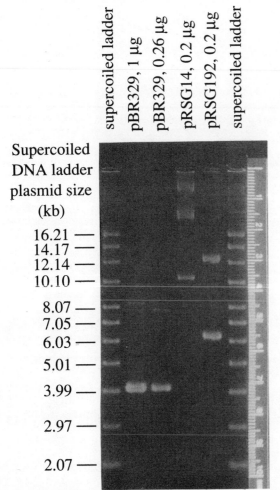

Figure 3-2. Agarose gel electrophoresis separation of undigested plasmid DNA.

A. C. Y. Chang and S. N. Cohen, *J. Bacteriol.* **134**, 1141 (1978).

L. Covarrubias and F. Bolivar, *Gene* **17**, 79 (1982).

M. Kahn, R. Kolter, C. Thomas, D. Figurski, K. Meyer, E. Remaut, and D. Helinski, *Methods Enzymol.* **68**, 268 (1979).

J. E. L. Larsen, K. Gerdes, J. Light, and L. Molin, *Gene* **28**, 45 (1984).

J. Norrander, T. Kempe, and J. Messing, *Gene* **26**, 101 (1983).

J. B. Sutcliff, *Cold Spring Harbor Symp. Quant. Biol.* **43**, 77 (1979).

S. Tabor and C. C. Richardson, *Proc. Natl. Acad. Sci. U.S.A.* **82**, 1074 (1985).

B. E. Uhlin, S. Molin, P. Gustafsson, and K. Nordström, *Gene* **6**, 91 (1979).

J. Vieira and J. Messing, *Gene* **19,** 259 (1982).

C. Yanisch-Perron, J. Vieira, and J. Messing, *Gene* **33**, 103 (1985).

INTRODUCTION OF DNA INTO CELLS

Recombinant DNA molecules can be introduced into *E. coli* cells in one of three ways: by transduction, by conjugation, or by transformation. **Transduction** involves the introduction of genetic information into a recipient cell by a bacteriophage. The most common vectors introduced by transduction are derived from bacteriophage, M13 or lambda (λ). M13 bacteriophage contain single-stranded circular molecules that replicate via a double-stranded intermediate. Vectors derived from M13 are frequently used in conjunction with dideoxynucleotide-chain termination DNA sequencing (see Lab VII).

Lambda vectors fall into two classes. The first class is specifically designed for cloning large fragments of DNA and is, therefore, useful for constructing clone banks. See Frischauf (1987) for construction and characterization of a genomic library in phage λ. A second class of vectors such as λgt10 and λgt11 (Huynh *et al.*, 1985; Jendrisak *et al.*, 1987) is designed for cloning cDNA (see Lab IX). Commercially available *in vitro* packaging systems allow the insertion into λ heads of foreign DNA. The only requirement is that such DNA contains λ *cos* sites spaced between 38 and 51 kb apart. Forty % of the λ genome can be replaced with foreign DNA before efficient plaque formation is lost, so the largest insert into λ vectors is between 20 and 25 kb (Kaiser and Murray, 1985). **Cosmids**, which are plasmids containing the λ *cos* site, can be packaged with inserts that are much larger (~45kb). Cosmids or λ-derived vectors that have been packaged into bacteriophage heads are introduced by transduction into *E. coli* cells, which is a very efficient means of moving recombinant DNA molecules into cells. See DiLella and Woo (1987) for techniques for cloning large segments of genomic DNA using cosmid vectors.

Conjugation involves plasmid-mediated transfer of DNA from one cell (donor) to another (recipient) through cell-to-cell contact. This is the chief means of introducing recombinant DNA molecules into several important Gram-negative organisms, such as *Agrobacterium tumefaciens* and *Rhizobium meliloti*. Vectors have been constructed from broad host range plasmids that facilitate transfer between *E. coli* and Gram-negative organisms. The most widely used approach for moving recombinant DNA molecules between *E. coli* and other Gram-negative bacteria involves the tripartite system employing the plasmids pRK290 and pRK2013 and derivatives of these plasmids (Ditta *et al.*, 1985). Cosmids have also been constructed from broad host range plasmids that can be transferred by transduction, conjugation, and transformation (Frey *et al.*, 1983; Gallie *et al.,* 1985).

Transformation results from the uptake of purified DNA by bacterial cells and is the most frequently used procedure for introducing recombinant DNA molecules into *E. coli*. Some species of bacteria, such as *Bacillus subtilis*, naturally take up DNA at a certain stage of growth called competence. *E. coli*, however, is not naturally competent at any stage of growth. Competence can be artifically induced in *E. coli* cells by treating them with calcium chloride prior to adding DNA (Cohen *et al.*, 1972). The Ca^{2+} destabilizes the cell membrane, and a calcium phosphate–DNA complex is formed, which adheres to the cell surface. The DNA is taken up during a heat-shock step when the cells are exposed briefly to a temperature of 42°C. Selection for cells containing transformed DNA is greatly enhanced if markers carried by the DNA are first expressed in the absence of selection by growing the cells for 1 to 2 hours in rich medium. The process by which *E. coli* cells are transformed is similar to the process used to transform eukaryotic cells (Chen and Okayama, 1988) in which a calcium phosphate–DNA complex is formed that adheres to the cell surface and is taken up by the cells after treating with cell wall-degrading enzymes, trypsin, or polyethylene glycol.

Choosing the proper genetic background for the recipient is important. For example, if transforming DNA is not methylated at *Eco*K sequences because the DNA was isolated from an organism other than *E. coli* K12, then the *E. coli* recipient should be *hsdR⁻*, which has the phenotype r⁻m⁺ (restriction⁻, modification⁺), and is missing the type I *Eco*K restriction endonuclease activity. If transforming with linear DNA, then a *recD* mutant deficient in the Rec BCD exonuclease (Exo V) activity should be used. This activity is primarily responsible for the 100- to 1000-fold lower transformation frequency of linear DNA compared to circular DNA, and in the absence of this activity, linear and circular DNA transform with equal frequency.

The transformants recovered in this experiment will be used in Lab V to map the *tet* gene encoding resistance to tetracycline (Tc) and to map intergenic regions of pBR329. Coselection for genetic markers carried on the plasmid (Ap^r, Cm^r) and for Km^r carried by Tn5 will yield Ap^r, Cm^r, Km^r transformants containing pBR329 molecules with Tn5 inserted into them. By following the procedures described in Labs III, IV, and V, a similar approach can be used to map any gene cloned in a plasmid.

Production of Frozen Competent Cells

MATERIALS

- ❏ 50 mM CaCl$_2$ (sterile)
- ❏ 50 mM CaCl$_2$, 15% glycerol (sterile)

- ❏ Overnight culture in 5 ml of LB broth of an *E. coli* recipient strain such as LE392

- ❏ LB broth

- ❏ −70°C freezer

- ❏ Container of ice

- ❏ Sorvall SS-34 rotor (or equivalent)

- ❏ Sorvall RC5B centrifuge (or equivalent)

PROTOCOL

1. Dilute overnight culture 1:100 into LB broth. The volume of LB broth depends on the number of transformation reactions required; 5 ml of culture yields one transformation reaction containing 0.2 ml competent cells. Shake the culture at 37°C until the OD_{450} is 0.3–0.4. Chill cells on ice 10 minutes.

2. Centrifuge the cells for 5 minutes at 5000 rpm, 4°C. Resuspend the cell pellets in 50 mM $CaCl_2$ using 1/4 the original volume of cells. Let sit on ice for 15 minutes.

3. Centrifuge the cells for 5 minutes at 5000 rpm, 4°C. Resuspend the cell pellet in 50 mM $CaCl_2$, 15% glycerol using 1/25 of the original volume of cells you started with.

4. Transfer 0.2 ml of competent cells to individual sterile microcentrifuge tubes and label tubes with permanent ink. Store immediately at −70°C. Place tubes on ice just before you are ready to add DNA.

Note: If the competent cells are to be used the same day, resuspend the cells in 50 mM $CaCl_2$ at 1/25 the original volume. No loss in transformation frequency will occur if the cells are held on ice for up to 5 hours before adding DNA as described below in the procedure for Day 2.

DAY 2

Transformation of LE392 with pBR329 DNA Isolated from HB101::Tn5

MATERIALS

❐ Microcentrifuge tubes containing competent LE392 cells (200 μl/ tube) in ice bucket

❐ pBR329 DNA isolated from HB101::Tn5 in Lab III

❐ Container of ice

❐ Sterile M9 salts

❐ Sterile tubes for making dilutions

❐ LB broth

❐ LB agar plates containing 50 μg/ml each of ampicillin, chloramphenicol, and kanamycin

❐ LB agar plates containing 50 μg/ml of ampicillin

❐ Turntable

❐ Hockey stick (bent glass rod for spreading cells)

❐ Petri dish containing 95% ethanol

❐ 37°C and 42°C water baths

Recipient strain: *E. coli* K12 LE392 F⁻ *lacY galK2 galT22 metB1 trpR55 supE44 supF58 hsdR514* (r_k^-, m_k^+)

47

PROTOCOL

1. Add at least 1 μg (10–20 μl) of pBR329 DNA isolated from HB101::Tn5 (see gel to estimate concentration) to a tube containing 200 μl of LE392 competent cells. Add an equivalent amount of water to a second tube of competent cells for a transformation control. Mix tubes gently.

   Note: Normally, much lower amounts, e.g., 50 ng, of transforming DNA can be used, but in this experiment, selection will be for the rare pBR329 molecule carrying Tn5 as well as for pBR329.

2. Keep on ice for 1 hour.

3. Heat shock: place tubes at 42°C for 1.5 to 2 minutes, then move tubes back to ice.

4. Add 1.0 ml of LB broth, and incubate tubes at 37°C for 1.5 to 2 hours to allow expression of the antibiotic resistance markers.

5. For both the transformation mixture and the control containing competent cells and water, mix cells, then spread 10 μl of a 1:10 dilution (made by adding 0.1 ml of the transformation mixture to 0.9 ml of M9 salts) on an ampicillin-containing plate.

6. Centrifuge the remaining cells in the transformation mixture and the control for 30 seconds, resuspend the cells in 0.1 ml of M9 salts, and spread all of the cells on an LB plate containing Ap, Cm, Km. Label plates clearly with your name, date, plasmid used, and strain transformed, and incubate plates overnight at 37°C. Examine plates and count colonies as described in Lab V, Day 1.

   Note: Simultaneous selection for Ap, Cm, and Km will yield colonies containing pBR329 "mutagenized" with Tn5. The location of the transposon insertion will be determined by restriction enzyme mapping in Lab V.

REFERENCES

C. A. Chen and H. Okayama, *Biotechniques* **6**, 632 (1988).

S. N. Cohen, A. C. Y. Chang, and L. Hsu, *Proc. Natl. Acad. Sci. U.S.A.* **69**, 2110 (1972).

A. G. DiLella and S. L. C. Woo, *Methods Enzymol.* **152**, 199 (1987).

G. Ditta, T. Schmidhauser, E. Yakobson, P. Lu, X. Liang, D. R. Finlay, D. Guiney, and D. R. Helinski, *Plasmid* **13**, 149 (1985).

H. M. Frey, J. M. Bagdasarian, D. Feiss, F. Christopher, H. Franklin, and J. Deshusses, *Gene* **24**, 299 (1983).

A.-M. Frischauf, *Methods Enzymol.* **152**, 190 (1987).

D. R. Gallie, S. Novak, and C. I. Kado, *Plasmid* **14**, 171 (1985).

T. V. Huynh, R. A. Young, and R. W. Davis, *in* "DNA Cloning: A Practical Approach" (D. M. Glover, ed.), Vol. 1, p. 49. IRL Press, Washington D.C., 1985.

J. Jendrisak, R. A. Young, and J. D. Engel, *Methods Enzymol.* **152**, 359 (1987).

K. Kaiser and N. E. Murray, *In* "DNA Cloning: A Practical Approach" (D. M. Glover, ed.), Vol. 1, p. 1. IRL Press, Washington D.C., 1985.

TN*5* MUTAGENESIS OF PBR329

Transposable elements of bacteria are distinct segments of DNA that have the capacity of transposing from site to site within a replicon or between replicons. The process by which they transpose does not involve homologous recombination and is RecA-independent. Transposons are transposable elements that contain coding regions other than for transposition, for example, an antibiotic resistance gene. The most commonly used bacterial transposons are Tn*5* (Berg and Berg, 1983), Tn*10* (Kleckner, 1986), and mini-Mu (Casadaban and Cohen, 1979), a derivative of the transposable bacteriophage Mu. Transposons, especially Tn*10*, are used to provide a selectable marker near a gene of interest in the *E. coli* chromosome. Such a marker (e.g., Tc^r carried by Tn*10*) provides a way of moving that gene into different genetic backgrounds. Currently, many genetic manipulations of the *E. coli* chromosome involve moving transposon-linked genes from one strain to another via P1 transduction, where random pieces of *E. coli* chromosomal DNA are packaged into the P1 phage head and transferred to a recipient.

Tn*5* and mini-Mu are useful for introducing mutations into DNA cloned on plasmids. (See Engebrecht *et al.* [1983] for applications of these transposons for determining operon organization and the direction of transcription.) Transposon mutagenesis, although a less convenient method for determining the location of a specific gene in a cloned sequence as compared to the analysis of mutations and deletions of the region constructed with restriction enzymes, has the advantage that these transposon-induced mutations, which carry an antibiotic resistance gene, can be moved back into the host chromosome from which the cloned

DNA was derived. The application of this method was first described for introducing chromosomal mutations into a gene involved in nitrogen fixation (*nif*) of *Rhizobium meliloti* (Ruvkun and Ausubel, 1981). These workers selected for Tn*5* insertions into a cloned *nif* gene then moved the Tn*5* insertions from the plasmid to the *R. meliloti* chromosome. In *E. coli*, the selectable replacement of a wild-type gene with a mutated gene is frequently carried out by replacing the gene in the cloned sequence with a restriction fragment carrying an antibiotic resistance marker, then moving the mutated sequence into the chromosome by homologous recombination. The Ω interposons, carrying different antibiotic resistance genes flanked in inverted orientations by transcription and translation termination signals and by restriction enzyme sites, are most useful for such insertional mutagenesis experiments (Prentki and Krisch, 1984; Fellay *et al.*, 1987). A suitable host for transforming with linearized DNA carrying the mutated gene is a *recD⁻* mutant, which is recombinogenic and deficient in RecBCD exonuclease activity.

The transposon, Tn*5*, is composed of a central region of 2.7 kb with two inverted repeat sequences (IS*50*) of 1.5 kb at each end. The insertion sequences denoted IS*50*L (left) and IS*50*R (right) differ by a single nu-

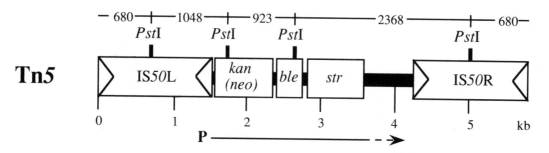

Figure 5-1. Genetic map of the transposon Tn*5*.

cleotide change at position 1442. IS50R codes for two proteins translated from a single reading frame, a transposase and an inhibitor of transposition (Johnson *et al.*, 1982; Isberg *et al.*, 1982). The single base change introduces an ochre mutation into this reading frame in IS50L, producing truncated proteins. The single base change also creates a promoter in IS50L for three antibiotic resistance genes in the central region: the *kan* gene (Beck *et al.*, 1982), coding for an aminoglycoside phosphotransferase that inactivates kanamycin and neomycin in bacteria and G418 in eukaryotic cells; a second gene, *ble*, conferring resistance to the bleomycin family of antibiotics (Collis and Hall, 1985); and the streptomycin resistance gene, *str*, that is not expressed in *E. coli* except when mutated (Mazodier *et al.*, 1986). The *kan* gene has been useful as a dominant selection marker for eukaryotic cells (Jimenez and Davies, 1980; Colbère-Garapin *et al.*, 1982) as well as bacteria, and was one of the first foreign genes to be expressed in plants (Bevan *et al.*, 1983). The complete nucleotide sequence of Tn5 has been determined (Auerswald *et al.*, 1981; Beck *et al.*, 1982; Mazodier *et al.*, 1985), and the map shown in Fig. 5-1 is based on this sequence.

In this lab exercise, transposon mutagenesis with Tn5 is used to define the extent of the tetracycline resistance gene and also the intergenic regions of the plasmid, pBR329. This plasmid contains genes specifying resistance to tetracycline, ampicillin, and chloramphenicol (see Fig. 3-1 in Lab III). Tn5 carries the kanamycin resistance gene. Your pBR329 DNA preparation was isolated from *E. coli* HB101::Tn5 (Km^r) in Lab III. With a transposition rate for Tn5 of 10^{-5}, a few pBR329 molecules that contain Tn5 will have been recovered in the plasmid miniprep isolated in Lab III. In Lab IV, you transformed *E. coli* strain LE392 with the plasmid DNA isolated in Lab III selecting for Ap^r, Cm^r, and Km^r. The Ap^r, Cm^r, and Km^r colonies isolated in Lab IV represent cells that were transformed with pBR329::Tn5 molecules. In this lab, the Ap^r, Cm^r, and

Km^r colonies will be examined for Tc^r and Tc^s. Those that are Tc^r will be used to define the intergenic regions of pBR329, and those that are Tc^s will be used to identify the *tet* gene. Plasmid DNA for each clone will be isolated, and restriction enzyme mapping of this DNA will locate the insertion site of Tn5.

DAY 1

Marker Screening: Divide Transformants into Tc^s and Tc^r Classes

MATERIALS

❏ LB agar plates containing 50 µg/ml of ampicillin, chloramphenicol, and kanamycin

❏ LB agar plate containing 15 µg/ml of tetracycline

PROTOCOL

1. Count the colonies on the plates from Lab IV, Day 2. Calculate the approximate transformation frequency (# Ap^r colonies/µg DNA) and the approximate transposition frequency into the *tet* gene and the intergenic regions (# Ap^r Cm^r Km^r colonies/# Ap^r colonies). Also, estimate the number of pBR329 molecules that contain Tn5 per µg in your DNA preparation.

2. In order to recover a purified clone, streak several different Ap^r, Cm^r, and Km^r colonies on Ap, Cm, and Km-containing agar plates so that isolated colonies will be recovered after incubation. Also, determine whether the colony is Tc^r or Tc^s by streaking part of the colony on a Tc-containing agar plate.

3. Incubate plates overnight at 37°C.

DAY 2

Purification of Tc^s and Tc^r Clones

MATERIALS

☐ 2 Ap, Km, Cm LB agar plates

PROTOCOL

Examine the plates streaked on Day 1 and pick one Tc^s and one Tc^r colony for further analysis. Streak cells from a single colony that grew on the Ap, Km, Cm-containing agar plate onto new Ap, Km, Cm-containing agar plates for isolation of plasmid DNA.

Note: It is important to purify the colonies by streaking as was done on Day 1, because cotransformation of pBR329 and the plasmid containing the transposon occurs frequently.

Isolation of Plasmid DNA by the Alkaline-Detergent Method and Determination of Recovery by Agarose Gel Electrophoresis

Refer to the procedure described in Lab III, Day 1 for rapid isolation of plamid DNA. For determining the plasmid DNA recovery, refer to the procedure given in Lab III, Day 2. Load 5 μl of the 50 μl in which the DNA pellet is dissolved on an agarose gel and separate by electrophoresis. After staining the gel with ethidium bromide, determine the amount of DNA in the plasmid preparation by comparing the fluorescence of your DNA band with the known amount of DNA run in another well.

Restriction Mapping of the Tn5 Inserts Using *Pst*I and *Eco*RI

Restriction enzyme digestion of DNA followed by gel electrophoresis permits the determination of the sizes of the DNA fragments produced by a particular restriction endonuclease. By analyzing fragments resulting from digestion of a particular piece of DNA, you can construct a restriction map. Mapping usually requires digestion with multiple enzymes, individually and in combination. In analyzing restriction fragments, it is important to keep in mind the following points:

1. Linear molecules that are not cut by a restriction enzyme run as a single band when separated by gel electrophoresis. Linear molecules that are cut once run as two bands, while those that are cut twice run as three bands (etc.).

2. Circular DNA will usually appear as two bands when uncut by a restriction enzyme (due to the presence of supercoiled and nicked forms of the molecule). The electrophoretic mobility of these molecules is aberrant relative to linear DNA (due to their conformation). When circular DNA is cut once, a linear fragment is produced, which migrates as a single band, and when cut twice, it runs as two bands (etc.).

3. The sum of the sizes of the fragments obtained after restriction digestion should equal the size of the initial DNA molecule. If this is not the case, more than a single fragment may be migrating as a single band. You may be able to determine this by observing the ethidium bromide fluorescence of the digest in question. Each band should contain an equivalent number of moles of each fragment. However, the larger fragments will fluoresce more brightly due to their greater mass. For example, a 4-kb DNA fragment should stain twice as intensely as a 2-kb DNA fragment. If the 2-kb band is of equal intensity to the 4-kb band, it is likely that the 2 kb band is composed of two different 2-kb fragments. See Fig. 5-2 for such an example.

Aside from digestion with multiple restriction enzymes, several other methods for restriction mapping are available. Perhaps the most powerful is the Smith–Birnstiel technique (1976) in which a DNA fragment with a single labeled end is subjected to partial digestion with a restriction enzyme. The digest is separated by electrophoresis alongside size markers. Autoradiography allows for the determination of the distance the various restriction sites are from the labeled end. This procedure may be applicable without resorting to radiolabeling, but success is dependent upon having a small number of cut sites. Another alternative procedure involves digestion of a DNA fragment with the enzyme Bal 31. This enzyme is a calcium-requiring double-stranded exonuclease. Bal 31 is added to the DNA for various time periods, followed by the addition of EGTA (to chelate calcium). The DNA is then digested with the restriction enzyme of choice. Electrophoresis will reveal which restriction fragments are located closest to the ends of the restriction map, because these will disappear first.

The enzymes *Pst*I (CTGCA↓G) and *Eco*RI (G↓AATTC) will be used to digest your plasmid DNA preparations of pBR329::Tn5. A single digest with *Pst*I, a single digest with *Eco*RI, and a double digest using both enzymes will provide you with enough information to identify where Tn5 is inserted in your Tc^r and Tc^s pBR329 clones. (See the restriction maps of pBR329 [Fig. 3-1] and Tn5 [Fig. 5-1]). There is only one *Eco*RI site and one *Pst*I site within pBR329, while there are no *Eco*RI sites and four *Pst*I sites in Tn5. The *Eco*RI/*Pst*I double digest of pBR329 yields two fragments, 1.4 and 2.75 kb. The strategy in determining Tn5 insertion sites involves first identifying which one of these fragments contains the Tn5 insertion. In the *Eco*RI/*Pst*I double digests of Tc^s and Tc^r pBR329::Tn5 clones shown in Fig. 5-2, the Tc^s clone has the 1.4-kb fragment but is missing the 2.75-kb fragment, while the Tc^r clone has the 2.75-kb fragment but is missing the 1.4-kb fragment. This indicates that the transposon has inserted into the 2.75-kb fragment of the Tc^s clone and into the 1.4-kb fragment of the Tc^r clone. To locate where in the *Eco*RI/*Pst*I fragment the transposon inserted, the single *Pst*I digestion pattern is compared to the *Eco*RI/*Pst*I double digestion pattern. Five fragments are expected in the *Pst*I digest because there are five *Pst*I sites in the pBR329::Tn*5* isolates. Because three of these fragments (0.92, 1.05, and 2.4 kb) are internal in the Tn*5* sequence, they will not be cut by *Eco*RI in the double digest. These three fragments can be seen in the *Pst*I and *Eco*RI/*Pst*I digests of both pBR329::Tn*5* isolates in Fig. 5-2. The *Pst*I fragment containing the *Eco*RI site will be missing in the double digest when compared to the *Pst*I digest and two new fragments will appear in its place as a result of cutting with *Eco*RI. The sizes of these two new fragments should add up to the size of the missing *Pst*I fragment. For the Tc[s] clone shown in Fig. 5-2, the *Pst*I fragment sizes are 3.0, 2.5, 2.4, 1.05, and 0.92 kb. When the *Eco*RI/*Pst*I double digest for the Tc[s] clone is examined, the 3.0 kb fragment is missing and two new fragments appear that are 1.6 and 1.4 kb in size. The 1.6-kb fragment indicates how far the closest Tn*5* *Pst*I site is from the *Eco*RI site as shown in Fig. 5-3. To find the site of insertion, the distance from the end of Tn*5* to the first *Pst*I site in Tn*5* (0.7 kb) must be taken into consideration. Subtracting 0.7 kb from 1.6 kb gives 0.9 kb, so Tn*5* inserted 0.9 kb to the right of the *Eco*RI site in the Tc[s] pBR329::Tn*5* isolate. The 2.5-kb *Pst*I fragment contains pBR329 sequence from the *Pst* I site to the point of insertion and Tn*5* sequence from the point of insertion to the first *Pst*I site in Tn*5*. Similar logic is used to deduce the Tn*5* insertion point for the Tc[r] isolate, which has *Pst*I fragments of 4.0, 2.4, 1.5, 1.05, and 0.92 kb in size. In the *Eco*RI/*Pst*I double digest, the 4.0-kb fragment is missing and two new ones, 2.75 and 1.3 kb, are present. The 1.3-kb fragment, after subtracting 0.7 kb for the distance from the end of Tn*5* to the first Tn*5* *Pst*I site, gives the Tn*5* insertion point, which is 0.6 kb from the pBR329 *Eco*RI site.

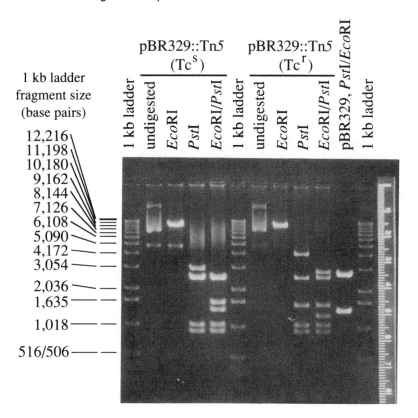

Figure 5-2. Agarose gel of restriction enzyme digests of two Tn*5* insertions into pBR329.

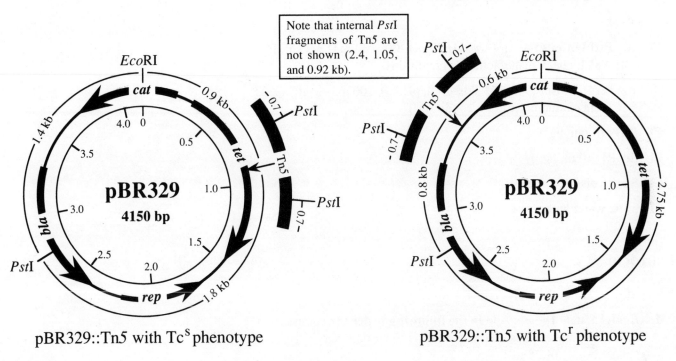

Note that internal *Pst*I fragments of Tn*5* are not shown (2.4, 1.05, and 0.92 kb).

pBR329::Tn*5* with Tc^s phenotype

pBR329::Tn*5* with Tc^r phenotype

Figure 5-3. Tn*5* insertion sites in pBR329 as deduced from restriction enzyme digestion patterns shown in Fig. 5-2.

61

MATERIALS

- ❐ 10x *Pst*I buffer (500 m*M* Tris–HCl, pH 8.0, 100 m*M* MgCl$_2$, 500 m*M* NaCl; check manufacturer's conditions)

- ❐ 10x *Eco*RI buffer (500 m*M* Tris–HCl, pH 8.0, 100 m*M* MgCl$_2$, 1 *M* NaCl; check manufacturer's conditions)

- ❐ *Pst*I (10 units/μl)

- ❐ *Eco*RI (10 units/μl)

- ❐ Sterile distilled water

- ❐ 37°C water bath

PROTOCOL

Set up digests for both the Tc^s and Tc^r clones:

1. ***Eco*RI Single Digest:** Add **in the following order** the components of the restriction digest reaction mixture to a microcentrifuge tube (final volume of 25 μl):

   _____ μl of sterile distilled water

   2.5 μl of 10x *Eco*RI buffer

   _____ μl of Tc^r or Tc^s DNA (~0.2–0.5 μg), estimate from gel

   _____ μl of *Eco*RI restriction endonuclease (3 units)   **MIX WELL**

2. ***Pst*I Single Digest:** Add **in the following order** the components of the restriction digest reaction mixture to a microcentrifuge tube (final volume is 25 μl):

   _____ μl of sterile distilled water

2.5 μl of 10x *Pst*I buffer

____ μl of Tc^r or Tc^s DNA (~0.5–1.0 μg), estimate from gel

____ μl of *Pst* I restriction endonuclease (5 units)   **MIX WELL**

3. ***Eco*RI and *Pst*I double digest:** Add **in the following order** the components of the restriction digest reaction mixture to a micro-centrifuge tube (final volume is 25 μl):

   ____ μl of sterile distilled water

   2.5 μl of 10x *Eco*RI buffer

   ____ μl of Tc^r or Tc^s DNA (at least1.0 μg), estimate from gel

   ____ μl of *Eco*RI restriction endonuclease (5 units)

   ____ μl of *Pst* I restriction endonuclease (5 units)   **MIX WELL**

Allow the six reaction mixtures to digest for either 2 hours or overnight at 37°C.

DAY 5

Agarose Gel of Plasmid DNA Restriction Endonuclease Digestions

MATERIALS

❏ 0.8% agarose gels, ~10 cm long

❏ Ethidium bromide

❏ 1-kb ladder from BRL

PROTOCOL

1. Mix each digest (25 µl) with 5 µl of loading buffer in the reaction tube. Load the samples into the wells of the agarose gel.

2. Load the 1-kb BRL ladder into a well on the same gel.

3. Add ethidium bromide at a final concentration of 0.5 µg/ml to the anode (+) buffer reservoir.

4. Connect the electrophoresis apparatus to the power supply and turn on the power.

5. Proceed with electrophoresis until the bromphenol blue dye has reached the bottom of the gel, which should take ~2 hours at 200 V or overnight at 20 V.

6. Photograph the gel and determine the distance migrated for each band in the 1-kb ladder lane. Prepare a semilog plot of the restriction fragment size (plot kb on the log scale) against the distance migrated (on the linear scale). Use this plot to calculate the size of restriction fragments seen in your digestions of plasmid DNA.

7. Calculate the Tn5 insertion sites in the Tc^s and Tc^r clones by constructing *PstI–Eco*RI restriction maps of the two plasmids using the restriction fragment sizes in the single and double digests and the maps of Tn5 and pBR329.

REFERENCES

E.-A. Auerswald, G. Ludwig, and H. Schaller, *Cold Spring Harbor Symp. Quant. Biol.* **45**, 107 (1981).

E. Beck, G. Ludwig, E. A. Auerswald, B. Reiss, and H. Schaller, *Gene* **19**, 327 (1982).

D. E. Berg and C. M. Berg, *Bio/Technology* **1**, 417 (1983).

M. W. Bevan, R. B. Flavell, and M.-D. Chilton, *Nature* **304,** 184 (1983).

M. J. Casadaban and S. N. Cohen, *Proc. Natl. Acad. Sci. U.S.A.* **76,** 4530 (1979).

F. Colbère-Garapin,F. Horodniceanu, P. Kourilsky, and A.-C. Garapin, *J. Mol. Biol.* **150,** 1 (1982).

C. M. Collis and R. M. Hall, *Plasmid* **14**, 143 (1985).

J. Engebrecht, K. Nealson, and M. Silverman, *Cell* **32**, 773 (1983).

R. Fellay, J. Frey, and H. Krisch, *Gene* **52**, 147 (1987).

R. R. Isberg, A. L. Lazaar, and M. Syvanen, *Cell* **23**, 191 (1982).

A. Jimenez and J. Davies, *Nature* (London) **287**, 689 (1980).

R. C. Johnson, J. C. Yin, and W. S. Reznikoff, *Cell* **30**, 873 (1982).

N. Kleckner, *in* "Regulation of Gene Expression" (I. R. Booth and C. F. Higgins, eds.), p. 221. Cambridge University Press, Cambridge, England, 1986.

P. Mazodier, P. Cossart, E. Giraud, and F. Gasser, *Nucleic Acids Res.* **13**, 195 (1985).

P. Mazodier, O. Genilloud, E. Giraud, and F. Gasser, *Mol. Gen. Genet.* **204**, 404 (1986).

P. Prentki and H. M. Krisch, *Gene* **29**, 303 (1984).

G. B. Ruvkun and F. M. Ausubel, *Nature* **289**, 85 (1981).

H. O. Smith and M. L. Birnstiel, *Nucleic Acids Res.* **3**, 2387 (1976) .

DNA CLONING IN M13

M13 bacteriophage are used as cloning vectors for DNA that is to be sequenced by the dideoxy chain termination method (Sanger *et al.*, 1980). Both double-stranded (ds) replicative form (RF) DNA in the infected cells and single-stranded (ss) + strand DNA packaged in the bacteriophage can be isolated in large amounts. Up to seven times the unit length of M13 can be packaged. However, DNA inserts greater than 2 kb in size in the multiple cloning site of the M13mp vectors are unstable (Messing, 1983), so these vectors are not useful for routine cloning experiments. The duplex RF of M13 can be isolated by any technique used for plasmid isolation. Isolation of ss DNA from the bacteriophage involves first precipitating the bacteriophage with polyethylene glycol (PEG), then removing the viral coat protein with organic solvents.

   M13 bacteriophage are male-specific bacteriophage in that they require the presence of F pili on the *E. coli* cell for infection. There are two ways to introduce M13 into a cell. The first way is to infect the cell with the whole virus; the F pilus is penetrated by the virus, the virus is stripped of its major coat protein in the cell membrane, and the viral circular ss DNA is converted into a circular ds RF DNA. The second way to introduce M13 into *E. coli* is to transform (transfect) competent cells with ss or ds M13 DNA. After replication of the bacteriophage DNA via a ds RF intermediate and a rolling circle mode of replication, the gene-5 protein–ss DNA complex moves to the membrane. There the coat protein molecules in the membrane bind to the DNA, displacing the gene-5 protein molecules. The mature particles are released without dissolution of the bacterial cell wall. M13 forms what appears to be plaques despite the lack of lysis of the host cell. The reason is that infected cells grow much more slowly than uninfected cells producing a less turbid region in the bacterial lawn.

The M13mp series of bacteriophage (Messing, 1983) contains a cluster of unique sites for cloning restriction fragments called the multiple cloning site (Table 6-1). These sites are engineered into the coding region of the α-peptide of β-galactosidase. The α-peptide contains approximately 15% of the amino-terminal end of β-galactosidase. Insertion of DNA into the multiple cloning site inactivates the α-peptide providing a means of selection for cloning DNA restriction fragments (Fig 6-1 and 6-2, Table 6-1). The *lac*Z gene, which encodes β-galactosidase, is defective or missing in *E. coli* strains [*e.g.* XL1- Blue (Bullock *et al.,* 1987) or JM109 (Yanisch-Perron *et al.,* 1985)] used for M13 infection. Strain JM109 has a *lac*Z chromosomal deletion, Δ(*lac*Z-*pro*AB), and on the resident F' plasmid there is a defective *lac*Z gene (*lac*ZΔM15), which codes for β-galactosidase lacking amino acids 11 through 41. In the absence of an α-complementing peptide, these strains contain an inactive dimer of β-galactosidase (Fig. 6-1B). Presence of an M13mp vector in these strains will permit the production of active β-galactosidase by interaction between α-peptide and the peptide produced from *lac*ZΔM15 (Fig. 6-1C). This phenomenon is called α-complementation (Langley *et al.,* 1975) and is detected by inducing the *lac* promoter with the gratuitous inducer, isopropylthiogalactoside (IPTG), and plating in the presence of the chromogenic substrate, 5-bromo-4-chloro-3-indolyl-β-D-galactoside (X-gal). An active tetrameric β-galactosidase is formed from the two polypeptides (Fig. 6-1C) and hydrolyzes X-gal yielding a blue precipitate. Bacteriophage that contain inserts of foreign DNA will not produce a functional α-peptide of β-galactosidase due to the disruption of the *lac*Zα gene on the M13 vector, and plaques will remain colorless

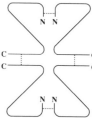

A. Biologically active tetrameric ß-galactosidase; non-covalent forces (......) maintain quaternary structure.

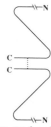

B. Inactive dimer produced in the deletion, *lac*ZΔM15, which removes amino acids 11-41 (\\).

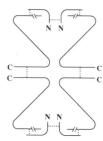

C. Dimer-dimer interaction restored by α-fragment (~15% of amino-terminus of ß-galactosidase) supplied by plasmid or M13 vectors.

Figure 6-1. β-galactosidase α-complementation (see Weinstock *et al.,* 1983).

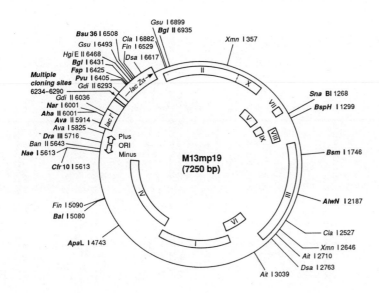

M13mp19 multiple cloning sites and primer binding region: 6201–6320

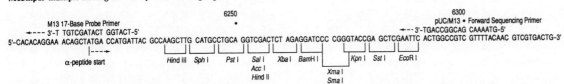

Figure 6-2. Restriction enzyme map of M13mp19 showing sequence of multiple cloning sites in the amino-terminal region of the α-peptide of ß-galactosidase (Messing, 1983; Yanisch-Perron *et al.,* 1985; sequence, van Wezenbeck, *et al.,* 1980). (Scale: 240 base pairs /cm; Boldface, single cleavage site.)

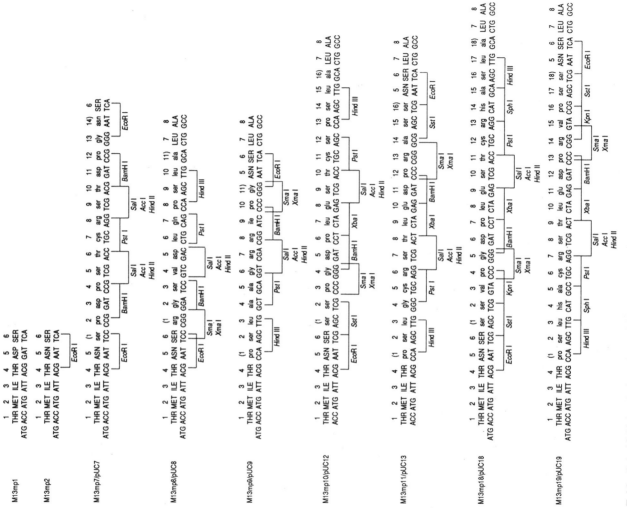

Table 6-1. Multiple cloning sites of M13mp bacteriophage and pUC plasmid vectors (Messing, 1983; Yanisch-Perron et al., 1985). Amino acid sequence is at the aminoterminus of α-peptide. Insertion of a restriction fragment into any of the restriction sites inactivates α-peptide.

in the presence of X-gal. Once the appropriate restriction fragment is cloned into the M13 vector, it can be sequenced using a primer (the universal primer) that hybridizes just upstream of the multiple cloning site. Fragments of several hundred bases can readily be sequenced as described in Lab VII. M13mp vectors, which come in pairs with the multiple cloning site in opposite orientations as shown in Table 6-1, are available for directional cloning and directed sequencing of both ends of a cloned fragment.

M13 Yields expected:

1. 3×10^{11} bacteriophage particles are equivalent to 1 µg of ss M13 DNA.

2. There are about 100 copies of RF and about 20 times as many bacteriophage excreted per cell.

3. The yield of a 10-ml culture grown to about 2×10^9 cells/ml should be ~12 µg of RF and 120 µg of ss DNA.

In the cloning experiment described here, you may substitute another restriction fragment in place of the one we chose to use. In this experiment, M13mp19 (Yanisch-Perron *et al.*, 1985) RF DNA will be isolated and used as a cloning vector for the 3.5-kb *Eco*RI fragment shown in Fig. 6-3 that encodes the chitobiase (*chb*) gene of the marine bacterium, *Vibrio harveyi* (Jannatipour *et al.*, 1987). This restriction fragment will be purified from an agarose gel by electrophoretically moving the fragment onto a piece of DEAE-cellulose membrane. Your purified M13mp19 RF DNA will be cut with *Eco*RI and treated with alkaline phosphatase to prevent religation of the vector. Optimal conditions for ligation of the *chb Eco*RI fragment and the vector treated with *Eco*RI and alkaline phosphatase will be established based on the concentrations of ends/ml of these fragments. In Lab VII, the nucleotide sequence of part of the 3.5-kb *Eco*RI fragment will be determined using the bacteriophage you construct in this lab.

Isolation of Restriction Fragment from an Agarose Gel

The 3.5-kb *Eco*RI fragment carrying the chitobiase (*chb*) gene will be isolated from pRSG192 (Fig. 6-3). The plasmid will be digested with

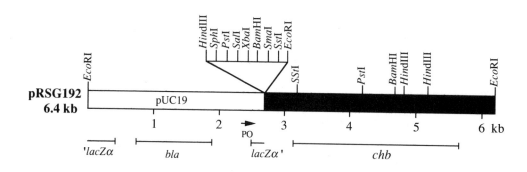

Figure 6-3. Restriction enzyme map of pRSG192 (from Jannatipour *et al.*, 1987).

*Eco*RI, the digest will be separated on an agarose gel, and the fragment containing the *chb* gene will be isolated from the agarose gel on a strip of DEAE-cellulose membrane.

A. *Eco*RI Restriction Digestion of pRSG192

MATERIALS

- ❐ pRSG192 (1.0 µg/µl)
- ❐ 10x *Eco*RI buffer
- ❐ *Eco*RI
- ❐ Sterile distilled water
- ❐ 37°C water bath
- ❐ Loading buffer
- ❐ BRL 1-kb ladder
- ❐ 10x TEA
- ❐ 0.8% agarose gel, 20 cm in length, poured with thick well makers so wells will hold 30 µl

PROTOCOL

1. Add **in the following order** the components of the restriction digest reaction mixture to a microcentrifuge tube (final volume of 25 µl):

    _____ µl of sterile distilled water

    2.5 µl of 10x *Eco*RI buffer

2 μl of pRSG192 (2 μg)

_____ μl of *Eco*RI restriction endonuclease (10 units)   **MIX WELL**

2. Allow the reaction to incubate for 2 hours at 37°C; then add 5 μl of loading buffer and load the total digest on an 0.8% agarose gel. Run the gel until the bromphenol blue dye is close to the bottom of the gel.

B. Restriction Fragment Isolation: Removal of the pRSG192 *Eco*RI Restriction Fragment Containing the *chb* Gene from Agarose

The 3.5-kb *Eco*RI fragment containing the *chb* gene will be recovered from the agarose gel using a band interception and blotting technique (Winberg and Hammarskjold, 1980; Dretzen *et al.*, 1981). This consists of first determining the position of the fragment by ethidium bromide staining and UV fluorescence, then cutting the gel just ahead of the band and inserting a small piece of NA45 DEAE-cellulose membrane in the cut. After running the fragment into the membrane, the fragment is removed from the membrane by eluting with a high salt buffer. The NA45 DEAE-cellulose membrane carries the functional group diethylaminoethyl in its protonated form and is used to immobilize anions. DNA, as you remember, is negatively charged. This membrane is stronger than DEAE paper and will not disintegrate when wet. The technique can be used to recover ss DNA, ds DNA, and RNA from agarose and polyacrylamide gels. For a description of other techniques for isolating DNA fragments from agarose or acrylamide gels, see Ogden and Adams (1987).

MATERIALS

☐ Agarose gel containing the *Eco*RI digestion of pRSG192

☐ Strips of NA45 DEAE-cellulose membrane (Schleicher & Schuell, Keene, NH)

☐ Sterile forceps

☐ Sterile scalpel

☐ Sterile microcentrifuge tubes

☐ Sterile high-salt NET buffer (high-salt NET: 1.0 M NaCl, 0.1 mM EDTA, 20 mM Tris–Cl, pH 8.0)

☐ 95% ethanol, in freezer

☐ 70% ethanol, at room temperature

☐ 68°C water bath

☐ Ethanol–dry ice bath

☐ UV light box

☐ Ethidium bromide

PROTOCOL

1. Stain the gel briefly for 5 minutes, then destain for 20 minutes. Examine gel on the UV light box but be careful to expose the DNA for as short a time as possible, because UV light will introduce nicks into DNA intercalated with ethidium bromide.

2. Using a scalpel or razor blade, make a slot just ahead of the band.

Cut a strip of NA45 DEAE membrane to fit into the slot, then wet the membrane with distilled water. Insert it into the slot. Maintain the sterility of the forceps and scalpel by dipping into ethanol and flaming.

3. Continue to run the gel at 200 V. After 15 minutes, examine whether the fragment is on the membrane with a UV light source.

4. Remove the strip and place in a microcentrifuge tube containing 300 µl of high-salt NET.

5. Spin the tube briefly in order to completely submerge the membrane.

6. Incubate for 1–1.5 hours at 68°C without shaking.

7. Determine whether the elution is complete by examining the membrane with the long-wave UV light.

8. Precipitate the DNA with 2.5 vol (750 µl) 95% ethanol (–20°C). Mix well and freeze in an ethanol–dry ice bath for 15–30 minutes. Centrifuge for 15 minutes. Decant the ethanol.

9. Wash the DNA pellet with 1.0 ml of 70% ethanol at room temperature, vortex briefly, spin for 5 minutes, and pour off the supernatant. Repeat two times.

10. Dry under vacuum and resuspend in 10 µl of sterile distilled water.

Estimation of Recovery of Restriction Fragment and Isolation of M13mp19 RF DNA

A. Estimation of Recovery of 3.5-kb *Eco*RI Restriction Fragment from Agarose Gel

MATERIALS

- ❏ Loading buffer
- ❏ BRL DNA Gel Markers, 1-kb ladder
- ❏ 10x TEA
- ❏ 0.8% agarose gel, 10 cm long
- ❏ Parafilm
- ❏ Plasmid control at known DNA concentration (pRSG192 cut with *Eco*RI)

PROTOCOL

1. Mix 2 µl of loading buffer with 5 µl of your restriction fragment on a piece of Parafilm.

2. Load sample plus loading buffer into well.

3. Load dilutions of a known amount of pRSG192 cut with *Eco*RI.

4. Load the 1-kb BRL ladder.

5. Connect the electrophoresis apparatus to the power supply, and run the gel until the bromphenol blue dye is at the bottom of the gel.

6. Stain the agarose gel with ethidium bromide and estimate the amount of DNA recovered by comparing with known amounts of pRSG192.

B. Isolation of M13mp19 RF DNA by the Alkaline-Detergent Method

Here you will isolate M13mp19 RF DNA, which is ds closed circular DNA. Because RF DNA is like plasmid DNA in structure, DNA isolation protocols are interchangeable. On Days 3, 4, and 5, the 3.5-kb *Eco*RI fragment carrying the *chb* gene will be cloned into the *Eco*RI site of the M13mp19 RF DNA.

MATERIALS

❒ Culture of XL1-Blue *recA1, endA1, gyrA96, thi, hsdR17, supE44, relA1, lacZ,* [F′, *proAB, lacI^q* ZΔM15, Tn*10*Tc^r] infected with M13mp19 bacteriophage.

❒ 2x YT plus tetracycline (Tc) medium (Bacto tryptone, 16 g/l; yeast extract, 10 g/l; NaCl, 5 g/l; Tc, 15 μg/ml; for plates add 1.5% agar and for soft agar add 0.6% agar. The soft agar after autoclaving is kept in a 55°C water bath prior to use.)

❒ IPTG, isopropylthio-β-D-galactoside: dissolve in water to prepare

100 m*M* solution and sterilize by filtration through a 0.2 μm Millipore filter. Store in refrigerator.

❏ X-gal, 5-bromo-4-chloro-3-indolyl-β-D-galactoside: dissolve in *N,N*-dimethylformamide (DMF) to make a 2% solution. Wrap aluminum foil around the tube to avoid damage by light, and keep refrigerated.

❏ Microcentrifuge.

❏ Materials for alkaline-detergent method of plasmid miniprep (Lab III, Day 1).

PROTOCOL

1. A single colony of XL1-Blue Tc^r grown on a 2x YT agar plate containing Tc is used to inoculate 5 ml of 2x YT plus 15 μg/ml Tc medium, and is allowed to grow overnight. A 1:10 dilution of the overnight culture of XL1-Blue is made into 250 ml of 2x YT medium containing 15 μg/ml Tc and is grown to an OD$_{450}$ of 0.3–0.5. The cells are infected with M13mp19 at an MOI (multiplicity of infection) between 1 and 10 (approximately 10^9 plaque-forming units, or pfu). Continue to shake cells at 37°C for 3–5 hours.

2. Isolate RF DNA following the alkaline-detergent method for plasmid DNA. Centrifuge 1.5 ml of culture in a microcentrifuge tube for 15 seconds, and decant the supernatant. Add another 1.5 ml of culture, centrifuge for 15 seconds, and decant the supernatant. Isolate M13mp19 DNA from the cell pellet as described in Lab III on p. 36.

3. The supernatant of the XL1-Blue culture infected with M13 can be

titered to determine the level of bacteriophage production in the culture. This supernatant can be frozen at −20°C and used to infect cells in the future. Expect ~10^{11} bacteriophage/ml. To determine the titer or bacteriophage concentration:

a. Streak XL1-Blue out on a 2× YT agar plate containing 15 µg/ml Tc and grow at 37°C overnight. Inoculate a single colony into 2 ml of 2× YT. Incubate for 5–12 hours at 37°C until the late log phase.

b. Make a number of 10-fold serial dilutions of the bacteriophage suspension.

c. Mix 0.1 ml from each dilution with 0.01 ml of 100 mM IPTG, 0.05 ml of 2% X-gal in DMF, 0.2 ml of the freshly grown XL1-Blue cells, and 3 ml of soft agar (45°C). Mix gently by rolling the tube in your hands and pour the soft agar mixture on 2× YT agar plates.

d. Let the soft agar harden and incubate the plates at 37°C for several hours. ("Plaques," or zones of retarded growth of infected cells can be seen after about 4 hours).

DAY 3

*Eco*RI Digestion of M13mp19 RF DNA and Treatment with Alkaline Phosphatase

The enzyme *Eco*RI will be used to digest your DNA preparation of M13mp19. This site is within the M13mp19 multiple cloning site and only appears once in M13mp19. The *Eco*RI fragment of pRSG192 encoding the *chb* gene will be inserted into the *Eco*RI site of M13mp19. After digesting M13mp19 with *Eco*RI, the bacteriophage DNA will be treated with alkaline phosphatase isolated from calf intestine (CIP). This enzyme will remove the 5' phosphate from the newly cleaved ends of the vector, thus preventing religation of the vector ends with one another during the ligation step.

MATERIALS

- ❏ 10× *Eco*RI buffer
- ❏ *Eco*RI
- ❏ Calf intestinal phosphatase (CIP)
- ❏ Sterile distilled water
- ❏ 37°C water bath

PROTOCOL

1. Add **in the following order** the components of the restriction digest reaction mixture to a microcentrifuge tube (final volume of 25 µl):

    _____ µl of sterile distilled water

    2.5 µl of 10× *Eco*RI buffer

    _____ µl of M13mp19 (~1.5 µg)

    _____ µl of *Eco*RI restriction endonuclease (10 units)

    _____ µl of CIP (0.5 units)   **MIX WELL**

2. Allow the reaction mixture to digest overnight at 37°C.

81

Removing the Phosphatase and Analysis of the *Eco*RI Digest of M13mp19 RF DNA

A. Removal of Phosphatase

MATERIALS

❏ Diethyl pyrocarbonate (DEPC) diluted 1:10 in 95% ethanol in an ice bucket

❏ 10x CIP stop buffer (100 m*M* Tris–Cl, pH 8.0, 3.0 *M* sodium acetate, 10 m*M* EDTA)

❏ Sterile distilled water

❏ TE buffer

❏ 70% ethanol, room temperature

❏ 100% ethanol, –20°C

❏ 68°C water bath

❏ Ethanol–dry ice bath

PROTOCOL

1. Add 65 µl of sterile water to your restriction digest of *M*13mp19 to bring the volume to 90 µl.

2. Add 10 µl of 10x CIP stop buffer to this same tube.

3. Add 1 µl of DEPC (diluted 1:10, final dilution = 1:1000); vortex

the DEPC just before pipetting.

4. Mix well and incubate at 68°C for 15 minutes.

5. Add 300 µl of 100% ethanol stored at –20°C.

6. Freeze in an ethanol–dry ice bath for 15 minutes.

7. Centrifuge 15 minutes at room temperature.

8. Remove the supernatant.

9. Wash the DNA pellet by adding ~1.0 ml of 70% ethanol (room temperature), vortex briefly, centrifuge for 5 minutes, and pour off the supernatant. Repeat two times.

10. Dry under vacuum and resuspend in 10 µl TE buffer.

B. Agarose Gel Electrophoresis of Restriction Digest of M13mp19 RF DNA

To be certain that you obtained good recovery of RF DNA in the mini-prep procedure and that the *Eco*RI digestion went to completion during CIP treatment, examine the digest using agarose gel electrophoresis before going on to ligation.

MATERIALS

❐ Loading buffer

❐ 10x TEA

❐ 0.8% agarose gel, 10 cm long

❐ 1-kb ladder

PROTOCOL

1. Mix 2 μl of loading buffer with 3 μl of your digested M13mp19 RF DNA.

2. Load your sample, a sample of your undigested M13 RF DNA, and dilutions of known amounts of M13 RF DNA. Also load the 1-kb ladder.

3. Run gel until bromphenol dye is at the bottom of the gel. After staining the gel with ethidium bromide, compare the amount of your undigested RF DNA with the known amount and estimate the yield. Examine the *Eco*RI digest to see if only the expected band corresponding to the size of M13mp19 is present by comparing to the known molecular weight standards in the 1-kb ladder. If the digest did not go to completion, a band migrating more slowly than the expected size of 7253 bp will appear, which will contain open circular RF molecules. Redigest the M13mp19 RF DNA if you do not see a single band because the open circular DNA will interfere with recovery of the desired recombinant product.

DAY 5

Ligation

The most significant factor in a ligation reaction is the concentration of ends; the higher the concentration of ends, the more likely the ends will find each other in solution. Ligation of DNA is a process in which bimolecular concatemerization competes with unimolecular cyclization. In other words, the two competing reactions in most ligation reactions are

(1) end-to-end joining of separate molecules and (2) joining of opposite ends of the same molecule.

For insertion of the pRSG192 3.5-kb *Eco*RI fragment into M13mp19 digested with *Eco*RI, the starting conditions should be such that the end-to-end joining of M13mp19 to the fragment is favored. Then, the conditions should change to favor the joining of opposite ends of the linear molecules.

The bimolecular reaction rate is dependent on the parameter, i, the concentration of reactive DNA termini in the reaction mixture. For self-complementary and blunt ends, the total concentration of ends/ml is

$$i = 2N_0 M \times 10^{-3} \text{ ends/ml} \qquad (6\text{-}1)$$

For nonidentical, cohesive termini, the equation is

$$i = N_0 M \times 10^{-3} \text{ ends/ml} \qquad (6\text{-}2)$$

where in both equations N_0 is Avogadro's number and M is the molar concentration of the DNA molecules.

The cyclization rate of linear DNA in the ligation reaction is governed by factor j, which is the effective concentration of one end in the vicinity of the other:

$$j = \left(\frac{3}{2\pi bl}\right)^{3/2} \text{ ends/ml} \qquad (6\text{-}3)$$

Here, l is the length of the DNA molecule, calculated by multiplying the number of base pairs by 3.4 Å, and b is the DNA statistical segment length. The Flory theory of polymers states that flexible, rod-shaped molecules such as DNA can be considered as a series of segments of length b, connected by swivels or universal joints. The value of b is dependent on salt, pH, and temperature, and most values for DNA center around 1200 ± 200 Å.

When $j = i$, any specific DNA end is equally likely to come into association with any other end, so the probabilities of linear and circular ligation products are equal, while $j/i < 1$ leads to increased formation of linear oligomers and $j/i > 1$ leads to increased formation of circular ligation products. You can see that as the concentration of ends decreases due to the formation of linear oligomers in a ligation reaction, conditions favorable to circular ligation increase.

For calculating j, it is convenient to use the following equation (Dugaiczyk *et al.*, 1975):

$$j_x = j_\lambda \left[\frac{mw_\lambda}{mw_x} \right]^{3/2} \text{ends/ml} \tag{6-4}$$

where j_λ is 3.6×10^{11}; mw_λ, the molecular weight of the λ genome (30.8×10^6 daltons); and mw_x is the molecular weight of the unknown DNA molecule.

Note that j is inversely proportional to molecular weight. The shorter the molecule, the greater the effective concentration of one end relative to the other in the same molecule. However, for very short molecules, the formula is no longer a reasonable estimate of j, because the DNA becomes rodlike in its behavior. Then j will be overestimated as will the tendency of DNA to cyclize.

The DNA concentration is related to j/i:

$$[\text{DNA}] = \frac{21.4}{(\text{kb})^{0.5}(j/i)} \text{ μg/ml} \tag{6-5}$$

Revie *et al.* (1988) provide some general rules that seem to work:

Table 6-2. Optimum Vector *j/i* for Different Insert: Vector *mw* Ratios

mw (Vector)/*mw* (Insert)	Optimum Vector j/i	[DNA] Vector [DNA] Insert
4.00	0.19	20
1.00	0.75	10
0.25	4.16	5

1. Ligation of an insert DNA fragment produced with a single restriction enzyme into a vector that has been treated with phosphatase where the vector cannot self-ligate: The efficiency of ligation of two ends of DNA where the phosphate group has been removed from one end but not from the other is very low, approximately 10% of the efficiency of phosphorylated ends. Self-ligation of the insert fragment that was not treated with phosphatase will proceed with normal efficiency. Revie *et al.* (1988) have calculated the optimum j/i for three different insert:vector molecular weight ratios where the vector has been treated with phosphatase and the insert has not (Table 6-2).

From this table, the optimum concentration of vector can be calculated from j/i in Eq. (6-5). Most vector concentrations give almost the same concentration of desired product, which is a circular molecule with one copy of vector and one copy of insert; however, the concentration of insert at the optimum concentration of vector gives the best yield of desired ligation product no matter what concentration of vector is used. Lower concentrations of vector may be warranted to avoid background problems resulting from incompletely dephosphorylated vectors.

For this exercise, calculate the j/i for the vector and insert, and estimate the ratio of vector to insert that you used. Where you did not recover enough DNA fragment to see on the gel, use 0.1 µg as the amount of DNA that you add to your ligation reaction. In the lab experiment with insertion of the pRSG192 3.5-kb *Eco*RI fragment into M13mp19 digested with *Eco*RI, the starting conditions should be such that the end-to-end joining of M13mp19 to the fragment is favored. Then, the conditions should change to favor the joining of opposite ends of the ligated molecules.

2. Forced directional cloning of an insert with two different restriction enzyme ends into a vector cut with the same two restriction enzymes: High cloning efficiency is achieved with this common scheme. Optimized ligation requires that vector and insert both be at low concentrations. This insures that the slow bimolecular associations between the reactant species will be followed by unimolecular cyclization. A high vector:insert ratio can be used to achieve virtually complete ligation of insert into circular products with one copy each of vector and insert. A low vector:insert ratio can be used to achieve virtually complete ligation of vector into the desired recombinants. Forced directional cloning is tolerant of poorly chosen ligation conditions, because neither the vector nor the insert is capable of self-ligation, and most of the multimeric product forms are not viable, such as head-to-head vector ligated products.

3. Cloning with λ DNA vectors where the aim is to produce linear concatemers rather than circular forms: Optimal results are obtained when two conditions are met: (a) the ratio of the concentration of restriction enzyme generated ends, i, for all three DNA species should be close to 1:1:1; and (b) the DNA concentration should be sufficiently high that $j/i < 1$.

4. Blunt-end ligation vs. cohesive end ligation: The temperature optimum for DNA ligase activity is 37°C. The T_m (midpoint melting temperature) of most cohesive ends generated by restriction endonucleases is below 15°C, so a compromise temperature of 15°C is used for ligation reactions containing cohesive ends. For blunt-end ligation reactions, use *j/i* values of less than 0.5. Blunt ends are ligated at 10% of the rate of sticky-ended DNA. Thermal stability of annealed cohesive ends is not a consideration with blunt-end ligations, so the temperature of blunt-end ligation reactions should approximate the T_m of the smallest fragment in the reaction, but not exceed 37°C.

A final ATP concentration of 0.5 mM is optimal for both cohesive and blunt-end ligations. In a reaction containing both cohesive and blunt ends where only ligation of the cohesive ends is desired, the cohesive ends will ligate at least 10 times faster than blunt ends so most of the products will result from ligation of cohesive ends. However, blunt-end ligation also can be preferentially inhibited by increasing the ATP concentration above 2.5 mM.

Ligation of *Eco*RI Digested M13mp19 RF DNA and the Purified pRSG192 *Eco*RI Fragment

MATERIALS

- ❏ 10x ligation buffer (0.5 M Tris–Cl, pH 7.8, 100 mM MgCl₂, 200 mM dithiothreitol, 10 mM ATP, 500 µg/ml bovine serum albumin)

- ❏ Your preparation of M13mp19 RF digested with *Eco*RI and treated with CIP

- ❏ Your preparation of purified 3.5-kb *Eco*RI fragment containing the *chb* gene of *V. harveyi*

- ❏ Container of ice

- ❏ T4 DNA ligase, 1 unit/µl

- ❏ 15°C water bath

PROTOCOL

1. Add the following reagents to two labeled 1.5-ml microcentrifuge tubes in the order listed (final volume = 20 μl):

Tube 1 (vector + fragment)
_____μl of sterile distilled water
2 μl of 10x ligation buffer
(~0.1 μg) of isolated *Eco*RI fragment
(~1.0 μg) of *Eco*RI digested and CIPed M13mp19
1 μl of T4 DNA ligase

Tube 2 (only vector)
_____ μl of sterile distilled water
2 μl of 10x ligation buffer
(~0.5 μg) *Eco*RI digested and CIPed M13mp19
1 μl of T4 DNA ligase

2. Allow the ligation reactions to incubate overnight at 15°C.

Transfection of XL1-Blue with Ligation Mixtures

MATERIALS

❑ One tube of competent XL1-Blue cells (600 μl/tube)

❑ Bottom agar plates (2x YT medium containing 15 μg/ml tetracycline, 15% agar); prewarm in 37°C incubator

❑ Tubes containing 3 ml top agar (2x YT medium containing 15 μg/ml tetracycline, 7.5% agar); melt the agar by boiling, and hold the tubes in a 50°C water bath.

❑ Overnight culture of XL1-Blue

❑ 2% X-gal in DMF

❑ Filter-sterilized 100 m*M* IPTG

❑ Water bath at 42°C

❑ Water bath at 50°C

❑ 37°C air incubator

❑ Container of ice

PROTOCOL

1. Divide the competent XL1-Blue cells into three tubes, 200 μl/tube. Keep competent cells on ice!

2. Label the tubes of competent cells #1, #2, and #3 and transfect with the following DNA preparations:

    a. To Tube #1, add contents of vector + fragment ligation reaction (Tube 1 of Day 5)

    b. To Tube #2, add contents of vector ligation reaction (Tube 2 of Day 5)

    c. To Tube #3, add 5 μl of your M13mp19 RF miniprep (see Day 2).

    Note: Mix gently after each DNA addition.

3. Leave samples on ice for 1 hour.

4. Heat shock tubes at 42°C for 1.5–2 minutes.

5. Put the tubes back on ice but let cells in these transfection reactions warm up before adding to soft agar (see step 7).

6. Add 200 μl of XL1-Blue overnight culture, 50 μl of 2% X-gal, and 20 μl of 100 m*M* IPTG to four of the tubes containing top agar. Make these additions just prior to use. Maintain these tubes at 50°C; otherwise the agar will harden.

7. Gently mix transfection reaction in Tube #1, then add 20 μl of Tube #1 to one top agar tube containing XL1-Blue cells, X-gal, and IPTG. Roll the tube in your hands to mix, and immediately pour onto a bottom agar plate appropriately labeled. Repeat with 200 μl of Tube #1, 200 μl of the transfection reaction in Tube #2, and 20 μl of the transfection reaction in Tube #3. Remember to mix the transfections right before diluting into the top agar because the cells will have settled out while sitting on ice.

8. Cool the plates on a level surface at room temperature for 10 minutes.

9. Incubate the plates at 37°C overnight.

DAY 7

Plaque Purification

In order to be certain that a clear plaque that you pick for analysis contains genetically identical bacteriophage, it is necessary to streak out each chosen plaque individually onto a new plate. One plaque from this plate will be chosen for growth and analysis of the recombinant bacteriophage DNA. Each plaque will also be examined for the ability to hydrolyze the artificial substrate for chitobiase, *p*-nitrophenyl-*N*-acetyl-ß-D-glucopyranoside (PNAG).

MATERIALS

❏ Plates containing your M13 transfections

❏ 5 bottom agar plates (2× YT medium containing 15 μg/ml tetracycline, 15% agar)

❏ 5 tubes containing 3 ml top agar, 7.5% agar in 2× YT+Tc medium. The agar is melted by boiling, and the tubes are then held in a 50°C water bath.

❏ Overnight culture of XL1-Blue cells

❏ Inoculating needles

❏ M13mp19 bacteriophage

❏ Filter-sterilized 2% X-gal (5-bromo-4-chloro-3-indolyl-ß-D-galactoside, the chromogenic substrate for ß-galactosidase) in DMF.

❏ Filter-sterilized 100 m*M* IPTG (isopropylthiogalactoside, gratuitious inducer of the *lac* promoter)

❏ Water bath at 42°C

❏ Water bath at 50°C

PROTOCOL

1. Examine the plates from the transfection experiment of Day 6. Were the XL1-Blue cells competent? What was the transfection frequency with the undigested M13mp19 DNA? Do you have colorless plaques?

2. Add 200 μl of XL1-Blue overnight culture, 50 μl of 2% X-gal, and 20 μl of 100 m*M* IPTG to four of the tubes containing top agar

which are in the 50°C water bath . Make these additions just prior to use. After addition of cells, maintain these tubes at 42°C (otherwise the agar will harden).

3. From the plate containing the M13+ restriction fragment transfection, choose one colorless plaque that is clearly separate from other plaques.

4. Using your sterile inoculating needle, transfer the plaque to two new bottom agar plates by making a single streak along the midline of these plates.

5. Pour the contents of two tubes of top agar containing XL1-Blue, X-gal, and IPTG onto the plates with the plaque streak (one tube/plate).

6. Repeat Steps 3, 4, and 5 for a second plaque.

7. Streak one bottom agar plate with M13mp19 bacteriophage, and pour the contents of one tube of top agar containing XL1-Blue, X-gal, and IPTG onto the plate.

8. Cool the five plates at room temperature for 10 minutes.

9. Store at 37°C overnight.

DAY 8

Isolation of Colorless Plaques that Contain the *chb* Gene and Growth of Recombinant Bacteriophage

MATERIALS

❑ Plates with recombinant plaques

❏ Spray bottle containing a solution of 10 mM PNAG (p. 92) dissolved in 100 mM sodium phosphate, pH 7.0

❏ 12 ml culture of XL1-Blue in 2x YT medium containing 15 µg/ml tetracycline (grown to mid-log phase)

❏ Inoculating needle

PROTOCOL

1. Using your sterile inoculating needle, select a single colorless plaque from one plate of each recombinant bacteriophage and inoculate 12 ml log phase cultures of XL1-Blue grown in 2x YT medium containing tetracycline (15 µg/ml).

2. Allow this tube to shake at 37°C for 3–5 hours. Isolate single-stranded recombinant bacteriophage DNA from the bacteriophage in the supernatant and recombinant RF DNA from the cells as described in Lab VII, Day 1 for determining *chb* insert orientation and nucleotide sequence of part of the *chb* gene.

3. Detection of chitobiase production by M13-producing cells in "plaques": Spray the other plates containing the recombinant bacteriophage and the plate containing M13mp19 with the PNAG solution. Observe after 5 minutes for yellow color. Those recombinant bacteriophage that produce clear plaques on X-gal-containing agar plates will now turn yellow if they contain the *chb* gene.

REFERENCES

W. Bullock, J. M. Fernandez, and J. M. Short, *Biotechniques* **5,** 376 (1987).

G. Dretzen, M. Bellard, P. Sassone-Corsi, and P. Chambon, *Anal. Biochem.* **112**, 295 (1981).

A. Dugaiczyk, H. Boyer, and H. M. Goodman, *J. Mol. Biol.* **96**, 171 (1975).

M. R. Jannatipour, W. Soto-Gil, L. C. Childers, and J. W. Zyskind, *J. Bacteriol.* **169**, 3785 (1987).

K. E. Langley, M. E. Villarejo, A. V. Fowler, P. J. Zamenhoff, and T. Zabin, *Proc. Natl. Acad. Sci. USA* **72**, 1254 (1975).

J. Messing, *Methods Enzymol.* **101**, 20 (1983).

R. C. Ogden and D. A. Adams, *Methods. Enzymol.* **152**, 61 (1987).

D. Revie, D. W. Smith, and T. W. Yee, *Nucleic Acids Res.* **16**, 10301 (1988).

F. Sanger, A. R. Coulson, B. G. Barrell, A. J. H. Smith, and B. A. Roe, *J. Mol. Biol.* **143**, 161 (1980).

G. M. Weinstock, M. L. Berman, and T. J. Silhavy, *In* "Gene Amplification and Analysis: Expression of Cloned Genes in Prokaryotic and Eukaryotic Cells," Vol. 3, Elsevier, New York, 1983.

P. N. G. F. van Wezenbeck, T. J. N. Hulsebos, and J. G. G. Schoenmakers, *Gene* **11**, 129 (1980).

G. Winberg and M-L. Hammarskjold, *Nucleic Acids Res.* **8**, 253 (1980).

C. Yanisch-Perron, J. Vieira, and J. Messing, *Gene* **33**, 103 (1985).

DNA SEQUENCING

Rapid determination of the sequence of recombinant DNA molecules has increased our knowledge of gene structure, expression, and function. Two types of DNA sequencing procedures are in common use today. These are the chemical degradation method of Maxam and Gilbert (1980) and the dideoxynucleotide-chain termination method of Sanger *et al.* (1980). The chemical degradation method is dependent upon end-labeling a single- or double-stranded DNA fragment. The labeled DNA fragment is divided into four aliquots and chemical reactions are performed so that a subset of the bases of one type (A,G,C,or T) are modified in each reaction. The DNA backbone is then cleaved at each modified residue by treatment with piperidine. Each aliquot thus results in a "nested set" of fragments, all labeled at a single end and terminating at the former location of a particular residue. These reactions are run in parallel lanes of a denaturing polyacrylamide gel. Frequently, multiple loadings of the samples are made so that different portions of the DNA sequence are maximally separated on the gel. The DNA fragment ladders are detected by autoradiography and the DNA sequence can subsequently be read by determining the order of the nucleotide residues from the labeled end. The Maxam and Gilbert technique is routinely used in conjunction with S1 nuclease mapping and primer extension experiments, where it is important to visualize the sequence of the digested or extended end-labeled fragment. The disadvantages of this technique include instability of some of the modification reactants—causing inconsistent results—and the hazardous nature of the chemical reagents (hydrazine, piperidine, dimethyl sulfate) used in this procedure.

The dideoxynucleotide-chain termination method requires annealing of a primer to a template strand that is to be sequenced. The hybrid is divided into four aliquots containing the four deoxynucleoside triphosphates (one of which is radioactively labeled) and a single dideoxynucleoside triphosphate. The primer is elongated by the addition of the Klenow fragment of DNA polymerase I (the large fragment that lacks 5' → 3' exonuclease activity), retroviral reverse transcriptase, or modified T7 DNA polymerase (Sequenase®). In each reaction, the elongating strand is occasionally terminated by the addition of a chain-terminating dideoxynucleotide. To prevent random termination of the chains, the reactions may be "chased" with a high concentration of dNTPs. The four reactions are subsequently separated on a denaturing polyacrylamide gel and the chain-terminated products are detected by autoradiography. The Sanger method is a reproducible technique and is preferred by many investigators for routine DNA sequencing. The dideoxynucleotide:deoxynucleotide ratios must be optimized so that they give the correct level of termination, i.e., if the ratio is too high, most molecules will terminate immediately; if it is too low, the dideoxynucleotides may never be incorporated into the growing DNA chain. These ratios are different, depending on the particular enzyme utilized for polymerization.

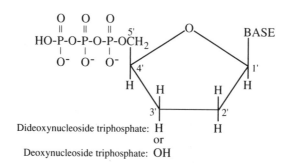

Figure 7-1 Deoxynucleoside triphosphate lacks a 2' hydroxyl group but has a 3' hydroxyl group. Dideoxynucleoside triphosphate lacks hydroxyl groups on both the 2' and 3' carbons.

Dideoxynucleotide-chain termination DNA sequencing is often utilized in conjunction with M13mp vectors that have been developed by Messing and collaborators (Yanisch-Perron *et al.*, 1985). M13 phage are single-stranded DNA viruses that infect male *E. coli* (F⁺, F', or Hfr strains). Their replication cycle involves a double-stranded intermediate that can be isolated from cells by a procedure similar to plasmid minipreps. The double-stranded form is used for cloning restriction fragments that will be sequenced. The single-stranded form isolated from the bacteriophage is used as a template for DNA sequencing. It should be noted that M13mp vectors are not useful for cloning DNA fragments of more than a few kilobases in length, since random deletions often occur. To obviate this problem, f1 phage/Col E1 plasmid vectors (phagemids; Dente *et al.*, 1983; see page 32), which do not exhibit nonspecific deletions in cloned sequences are increasingly being used.

In recent years, procedures for dideoxynucleotide sequencing of double-stranded DNA have been developed (Chen and Seeburg, 1985; Hattori and Sakaki, 1986). These involve alkaline denaturation of the template DNA followed by annealing of a primer and chain elongation. The universal primer can be utilized for this procedure when fragments are cloned into pUC plasmids that contain the same multiple cloning sites as M13mp vectors (see Table 6-1). Other recent advances include the use of low energy [³⁵S]-labeled dNTPs, which permit safer manipulation of the reactants and better resolution on autoradiographs. Further, the use of gradient gels or wedge-shaped gels permit analysis of more DNA sequence per gel lane. Resolution of longer DNA fragments is enhanced, since the smaller fragments are more closely spaced in the higher concentration of acrylamide or thicker portion of the gel.

Only several hundred bases of sequence can be determined from a single sequencing reaction. With the Maxam and Gilbert method, it is usually necessary to label and sequence overlapping restriction fragments to generate the entire sequence of a clone. For the chain termination DNA sequencing method, it is cumbersome to sequence regions of a cloned DNA fragment that are not located near the junction with the vector. One option is to subclone smaller fragments (either randomly or via directed cloning). A second method is to delete the terminal regions of the insert so as to be able to read further into the sequence (Dale *et al.,* 1985). (Kits for this purpose are available from a number of molecular biology supply companies.) Finally, one may chemically synthesize a primer complementary to the most distal sequence known and use this to sequence further into the fragment of interest.

In this series of protocols, you will prepare double-stranded and single-stranded phage DNA from M13-infected *E. coli.* You will use the double-stranded replicative form of the phage to map the orientation of the insert in the vector and the single-stranded DNA for dideoxynucleotide-chain termination DNA sequencing.

DAY 1

Isolation of Recombinant M13mp19 RF and ss DNA

MATERIALS

- ❐ Recombinant M13mp19-infected XL1-Blue cultures (see final steps of Lab VI, Day 8 for culture preparation information)

- ❐ 40% polyethylene glycol (PEG)

- ❐ 5 *M* sodium acetate

❏ 20 m*M* Tris–Cl, pH 7.2

❏ Phenol saturated with 1 *M* Tris–Cl, pH 8.0

❏ Phenol:chloroform (1:1), saturated with TNE

❏ Chloroform, saturated with TNE

❏ 3 *M* sodium acetate

❏ 100% ethanol at –20°C

❏ 70% ethanol at room temperature

❏ Materials for miniprep isolation of plasmid DNA (see Lab III, Day 3)

❏ Ethanol–dry ice bath

❏ Sorvall superspeed centrifuge and SS34 rotor

❏ 15 ml Corex centrifuge tubes

PROTOCOL

1. Centrifuge 12 ml of recombinant M13mp19-infected culture at 10,000 rpm (12,000 *g*) for 10 minutes in Sorvall centrifuge and decant supernatant to a 15-ml Corex tube.

2. Resuspend the pellet in 5 ml of TNE and transfer 1.5 ml to a microcentrifuge tube. Isolate RF (double-stranded, replicative form) DNA from the pellets following the instructions given previously for the miniprep isolation of plasmid DNA (see Lab III, Day 3), but **dissolve the final DNA pellet in 20 μl TE** buffer.

3. The 12 ml of supernatant from Step 1 contains the recombinant M13 bacteriophage, from which you will isolate the single-

stranded phage genome. Precipitate the phage by adding 670 μl of 40% PEG and 670 μl of 5 M sodium acetate to the 15-ml Corex tube.

4. Place on ice for 30 minutes.

5. Centrifuge for 15 minutes at 10,000 rpm and discard the supernatant.

6. Drain excess liquid by placing the Corex tube upside down on a paper towel for 10–30 minutes.

7. Wipe remaining liquid from around rim with a Kimwipe.

8. Resuspend the phage in 600 μl of 20 mM Tris–Cl, pH 7.2.

9. Extract with equal volumes of

   a. phenol saturated with 1 M Tris–Cl, pH 8.0, one time

   b. phenol:chloroform (1:1), two times

   c. Chloroform saturated with TNE, one time

10. Add 3 M sodium acetate to a final concentration of 0.3 M (~25μl).

11. Add 1 ml of 100% ethanol and mix gently.

12. Freeze in ethanol–dry ice bath for 20 minutes.

13. Centrifuge for 10 minutes.

14. Wash DNA pellet: Add ~1.0 ml of 70% ethanol, vortex briefly, spin for 5 minutes, and pour off supernatants. Repeat two times.

15. Dry pellet under vacuum to remove ethanol.

16. Dissolve the precipitated DNA in 25 μl of sterile distilled water. Mix well and let sit at room temperature for 10 minutes. Label tubes. Store at –20°C.

Restriction Digestion and Gel Electrophoresis of Recombinant Phage to Determine Orientation of Insert

MATERIALS

- ❏ 10x *Pst*I buffer (stored at -20°C)
- ❏ 10x *Eco*RI buffer (stored at -20°C)
- ❏ *Pst*I (stored at –20°C)
- ❏ *Eco*RI (stored at –20°C)
- ❏ Microcentrifuge tubes
- ❏ Sterile distilled water
- ❏ 37°C water bath
- ❏ Miniprep of recombinant RF DNA
- ❏ Miniprep of single-stranded M13 recombinant
- ❏ M13 single-stranded DNA (at known concentration)
- ❏ Control plasmid DNA (in this case, pRSG192 DNA at 0.1 µg/µl)
- ❏ 0.8% agarose gel (12 cm)
- ❏ TEA gel buffer
- ❏ Gel loading buffer
- ❏ Electrophoresis chamber

PROTOCOL

1. Add **in the following order** the components of the restriction digestions to two microcentrifuge tubes (final volume for each is 25 μl).

Digest 1 (miniprep)

___μl of sterile distilled water

2.5 μl of 10x *Eco*RI buffer

10 μl of recombinant RF DNA (~0.5-1.0 μg)

___μl of *Pst*I restriction endonuclease (10 units)

Digest 2 (pRSG192)

___μl of sterile distilled water

2.5 μl of 10x *Pst*I buffer

5 μl of pRSG192 (0.5 μg)

___μl of *Pst*I (10 units)

___μl of *Eco*RI (10 units)

2. Mix well after adding all components, and allow the reaction mixture to digest 2 hours at 37°C.

3. Estimate how much RF DNA you would have recovered from the culture if there were 100 RF molecules/cell, recovery was 100%, the size of the recombinant M13 RF molecule is 10.8 kb (3.5 kb + 7.3 kb), and there were 10^9 cells/ml.

4. Add 5 μl of loading buffer to the *Pst*I digests of the recombinant RF DNA and to the *Pst*I and *Eco*RI digested pRSG192.

5. Load a 0.8% agarose gel with the two samples and run the gel overnight at 15 V.

Estimating the Recovery of ss DNA Using Agarose Gel Electrophoresis

MATERIALS

- ☐ 0.8% agarose gel (12 cm)
- ☐ TEA gel buffer
- ☐ Gel loading buffer
- ☐ Electrophoresis chamber
- ☐ M13 ss DNA

PROTOCOL

1. Mix 3 μl of your ss DNA preparation with 2 μl of gel loading buffer and load them on a 0.8% agarose gel. Electrophoresis should be performed for 1 hour at 150 V.

2. Include known amounts of M13 ss DNA on the same gel as controls.

3. Stain gel with ethidium bromide as previously described (see Lab II), and estimate the amount of ss DNA in your minipreps.

DAY 3

Staining Gel of *Pst*I Restriction Fragments

PROTOCOL

1. Stain the gel containing the *Pst*I digests of the *Eco*RI *chb* fragment cloned into M13mp19 (see Lab II for ethidum bromide staining

procedure and Lab VI concerning cloning and restriction map of the *chb* fragment).

2. Determine the orientation of *Eco*RI insert and choose one phage with one orientation and one with the other for dideoxy sequencing.

3. Determine that there is enough ss DNA for sequencing these four DNAs by examining the gel from the previous lab containing ss DNA preparations (a minimum of 0.5 µg of DNA is required).

Template Annealing and Dideoxy Sequencing

You will be determining the nucleotide sequence of each end of the *Eco*RI fragment encoding the *chb* gene. One orientation will have one end of the *Eco*RI fragment closest to the M13mp19 sequence that hybridizes to the 17-base universal primer and the other orientation will have the other end closest.

Materials

- ❐ Single-stranded DNA preps with each orientation of the *Eco*RI fragment

- ❐ 5x sequencing buffer (200 m*M* Tris–Cl, pH 7.5, 50 m*M* MgCl$_2$, 250 m*M* NaCl)

- ❐ 17-base universal sequencing primer at ~3 ng/µl

- ❐ Klenow fragment of DNA polymerase I at 1 unit/µl

- ❐ 1x extension mix (1.5 µ*M* dCTP, 1.5 µ*M* dGTP, 1.5 µ*M* CTP). Prepare by diluting 5x extension mix with sterile distilled water prior to use. dATP is not included because it is the radiolabeled nucleotide and will be added separately.

Note: pH of stock nucleotide solutions should be adjusted to 7.0 using 0.05 *M* Trisma base and pH paper.

❏ 0.1 *M* dithiothreitol

❏ [^{35}S]dATP (1000 Ci/mmol, 10mCi/ml)

❏ Termination mixes (pH of stock solutions should be adjusted to 7.0 using 0.05 *M* Trisma base and pH paper)

A: 300 μ*M* ddATP, 25 μ*M* dATP, 250 μ*M* dCTP, 250 μ*M* dGTP, 250 μ*M* TTP

C: 100 μ*M* ddCTP, 25 μ*M* dCTP, 250 μ*M* dATP, 250 μ*M* dGTP, 250 μ*M* TTP

G: 150 μ*M* ddGTP, 25 μ*M* dGTP, 250 μ*M* dATP, 250 μ*M* dCTP, 250 μ*M* TTP

T: 500 μ*M* ddTTP, 25 μ*M* TTP, 250 μ*M* dATP, 250 μ*M* dCTP, 250 μ*M* dGTP

❏ Stop solution (95% formamide, 20 m*M* Na$_2$EDTA, 0.05% [w/v] bromphenol blue, 0.05% [w/v] xylene cyanol FF)

❏ Microcentrifuge tubes

❏ Microcentrifuge

❏ 68°C water bath

❏ 37°C water bath

❏ Gloves

PROTOCOL

Template–Primer Annealing

1. Mix the following in a 1.5-ml microcentrifuge tube:

    _____ μl of M13mp19 recombinant (+) ss DNA (1 μg)

    2 μl of sequencing buffer (5x)

    _____ μl of M13 17-base universal primer (3 ng, 0.5 pmol)

    _____ μl of sterile distilled water (to 10 μl)

2. Close the tube, mix, and centrifuge briefly.

3. Heat the tube for 3 minutes at 68°C, then cool slowly (30 minutes) to 30°C or room temperature. The tube can be placed at room temperature or in a heating block set at 30°C.

4. During the cooling period, label four tubes A, C, G, or T. Add 2.5 μl of the appropriate termination mix. Prewarm to 37°C.

Labeling-Extension Reaction

5. Centrifuge the tube briefly.

6. The following additions to the 10 μl of annealed template-primer should be made in the order below when you are ready to add [^{35}S]dATP and the Klenow fragment:

    2 μl of extension mix (1x)

    1 μl of 0.1 M dithiothreitol

    2 μl of Klenow fragment of DNA polymerase I (1 unit/μl)

    0.5 μl of [^{35}S]dATP —start timer with this addition

7. Mix, centrifuge briefly, then incubate the tube at room temperature for 5 minutes.

Termination Reaction

8. Transfer 3.5 µl aliquots of the labeling-extension reaction to the four prewarmed tubes labeled A, C, G, and T.

9. Incubate at 37°C for 15 minutes.

10. Add 4 µl of the stop solution.

11. Freeze samples at –20°C until you are ready for gel electrophoresis.

CAUTION: Observe proper radiation safety precautions (lab coat, gloves, eye protection, absorbent paper). Discard radioactive material only in radioactive waste containers.

DAY 4

Electrophoretic Separation of Sequencing Reactions

Gel electrophoresis is the method of choice for separating DNA sequencing reactions and numerous sequencing gel apparatuses are commercially available. These consist of two glass plates separated by thin (0.25–1 mm) spacers and a vertical electrophoresis chamber. One of the glass plates is first treated with siliconizing solution or Pam nonstick spray to allow easier removal following electrophoresis. The plates are assembled and sealed at the bottom and edges with water-resistant tape, a gasket system, and/or rubber cement. The polyacrylamide solution is prepared by the addition of TEMED and ammonium persulfate to a 6–20% acrylamide solution (containing 1/30 the concentration of bis-acrylamide). The gel solution also contains a high concentration of urea (to enhance DNA denaturation) as well as Tris-borate, EDTA running buffer. The gel solution is slowly poured into the assembled gel plates, taking care to prevent the trapping of air bubbles. A well-former (comb) is inserted at the top of the gel. (Currently many researchers form a flat top on their gel and use a "shark's-tooth" comb to form their wells following polymerization.) Prior to electrophoresis, the bottom tape or gasket is removed, the gel is attached to the electrophoresis chamber, and the chamber is filled with buffer. The wells are rinsed with buffer (using a syringe) to remove unpolymerized acrylamide, and the gel is pre-run to heat up the gel plates to approximately 45°C (this aids in DNA denaturation). The samples are denatured by heating in a formamide buffer and are loaded in the gel wells. Following electrophoresis, the gel plates are disassembled, and the gel is fixed and dried. Sequencing termination products are detected by autoradiography (see Bonner, 1987).

Caution: When pouring acrylamide gels, wear a lab coat and gloves. Rinse skin immediately after any contact with acrylamide solution. Acrylamide is a neurotoxin and may have mutagenic properties.

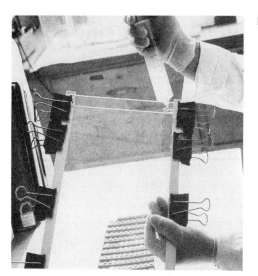

Figure 7-2. Pouring a sequencing gel.

 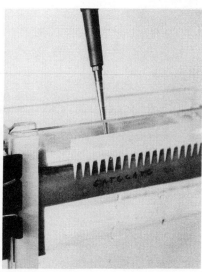

Figure 7-3. Loading a sequencing gel. Note the "shark's-tooth" comb that serves as a well-maker.

MATERIALS

☐ 2x TBE-sequencing gel buffer with urea and 8% polyacrylamide (one liter contains 200 ml of 10x TBE-sequencing gel buffer, 500 g of urea, 80 g of acrylamide, 2.66 g of bis-acrylamide)

☐ TEMED

☐ 10% ammonium persulfate

☐ Face mask for filtering dust when weighing out acrylamide

☐ 2x TBE sequencing gel buffer [one liter of 10x contains 121.1 g of Tris base, 53 g of boric acid, 7.44 g of EDTA (free acid)]

☐ Sequencing gel plates, spacers, and combs

☐ Sequencing gel electrophoresis apparatus and power supply

Warning: When making solutions containing acrylamide, care must be taken to avoid contact with skin and breathing acrylamide powder. When weighing out acrylamide, wear a lab coat, gloves and a face mask to filter dust.

- ❐ 75°C water bath
- ❐ Microcentrifuge
- ❐ Gel fix (10% acetic acid, 10% methanol)
- ❐ Filter paper
- ❐ Gel dryer
- ❐ Autoradiography folder
- ❐ X-Ray film
- ❐ Gloves

PROTOCOL

1. A polyacrylamide Tris-borate-urea sequencing gel should be poured as described above. The gel is prepared from 150 ml of 2x TBE-sequencing gel buffer with urea and 8% polyacrylamide, 1 ml of 10% ammonium persulfate, and 25 μl of TEMED.

   Caution: Wear gloves throughout!

2. Following polymerization of the gel, remove the comb and replace it with a shark's-tooth comb. Assemble the gel onto the gel apparatus, fill the buffer chambers, and rinse the wells with a syringe to remove any unpolymerized acrylamide.

3. Incubate A, C, G, and T samples at 75°C for 2 minutes.

4. Centrifuge briefly.

5. Load 3 μl of the samples on the sequencing gel in the order GATGCATC.

6. Run the gel for 4 hours at 2500 V.

7. Remove the gel from the apparatus. Separate the two plates by inserting a spatula and carefully lifting away the siliconized plate.

8. Fix for 15 minutes.

9. Dry gel on a filter paper backing.

10. In a darkroom, expose the gel to X-ray film by placing in a film folder and clamping to prevent light leaks. Store at room temperature until you are ready to develop (1–5 days).

DAY 5

Developing Autoradiogram and Reading DNA Sequence

MATERIALS

- ❏ Developer (e.g., Kodak GBX)
- ❏ Fixer   (e.g., Kodak GBX)
- ❏ 3 developing trays
- ❏ Water

PROTOCOL

Developing the Film

1. In a darkroom, remove film from folder and place in X-ray developer for 2–3 minutes. If the film appears to be overexposed, developing can be done for a shorter period.

2. Transfer the film to water for 30 seconds to wash off excess developer.

113

3. Fix in X-ray fixer for 3 minutes.

4. Wash in water for 2 minutes.

5. Air-dry the film.

6. Label the autoradiograph to indicate which sequencing reactions and which samples are in each lane.

7. Read the DNA sequence beginning at the bottom of the gel (smallest fragment, closest to the primer). The sample loading order should help you determine the precise sequence since each termination reaction is in an adjacent lane to each of the other three termination reactions.

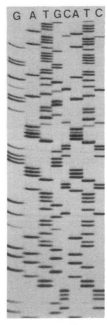

Figure 7-4. Autoradiograph of DNA sequence produced by the dideoxynucleotide-chain termination method.

REFERENCES

W. M. Bonner, *Methods Enzymol.* **152**, 55 (1987).

E.Y. Chen and P. H. Seeburg, *DNA* **4**, 165 (1985).

R. M. K. Dale, B. A. McClure, and J. P. Houchins, *Plasmid* **4**, 31 (1985).

L. Dente, G. Cesareni, and R. Cortese, *Nucleic Acids Res.* **11**, 1645 (1983).

M. Hattori and Y. Sakaki, *Anal. Biochem.* **152**, 232 (1986).

A.M. Maxam and W. Gilbert, *Methods Enzymol.* **65**, 499 (1980).

F. Sanger, A. R. Coulson, B. G. Barrell, A. J. H. Smith, and B. A. Roe, *J. Mol. Biol.* **143**, 161 (1980).

C. Yanisch-Perron, J. Vieira, and J. Messing, *Gene* **33**, 103 (1985).

DNA GEL BLOTTING, PROBE PREPARATION, HYBRIDIZATION, AND HYBRID DETECTION

In 1975, Southern described a technique for transferring electrophoretically separated DNA fragments from agarose gels to nitrocellulose (NC) paper. This procedure is useful for detecting homology between nucleic acids. Denatured DNA bound to an NC filter can be probed with radioactively labeled nucleic acids. After hybridization and washing, the filter is exposed to X-ray film, to detect which bound DNA bands have hybridized to the radioactive probe. Biotinylated DNA can be used in place of a radioactive probe, and the hybridized DNA can be detected using avidin, which binds biotin. The avidin is conjugated to an assayable enzyme, such as alkaline phosphatase. The hybrid is detected using a chromogenic substrate (dye) that is only colored after alkaline phosphatase treatment.

The transfer procedure devised by Southern involves denaturing the DNA duplexes within the gel by treatment with base. (Single-stranded DNA binds to NC, while double-stranded DNA does not.) The gel is then neutralized and transferred to a filter paper wick suspended on a plate above a basin of high-salt buffer. A piece of wet NC paper is placed over the gel, followed by two or three pieces of filter paper. A stack of paper towels is placed on top of the filter paper and a weight is placed on top of this. The wicks serve to draw the solution from the basin to the gel and the paper towels draw the liquid through the gel, transferring the DNA to the NC paper in the process. After the transfer is complete, the NC paper is rinsed briefly in a low-salt solution and baked for 2 hours in a vacuum oven. This process binds the DNA to the paper. In recent years, a number of procedures substituting nylon filters for NC filters have been devised, as nylon filters are less brittle than NC paper. Due to their superior strength, nylon filters are amenable to repeated removal of bound radioactive probe and rehybridization with different probes.

The following transfer technique is an adaptation of Southern's. It allows the rapid bidirectional transfer of DNA from gel to paper. Thus from one gel, two blots can be prepared (Smith and Summers, 1980). You will be able to detect the hybridization using autoradiography or a dye reaction, and this experiment should allow you to evaluate the relative sensitivity of the two techniques. The hybridization portion of this procedure can be performed simultaneously on filters prepared from a λ phage library (see Lab IX).

DAY 0

Agarose Gel Electrophoresis

Prepare an agarose gel and separate restriction fragments by electrophoresis. Some of these fragments should be capable of hybridizing to the probe you prepare in Day 2. For example, you could probe *V. harveyi* chromosomal DNA with the *Eco*RI fragment containing the *chb* gene. In that case, you would digest *V. harveyi* as *Eco*RI and separate the fragments on an agarose gel by electrophoresis. You would then label pRSG192 by nick translation in Day 2.

Gel Blotting

MATERIALS

- ❐ Denaturation solution (0.5 M NaOH, 1.5 M NaCl)
- ❐ Neutralizing solution (1 M ammonium acetate, 0.2 M NaOH)
- ❐ Distilled water
- ❐ Whatman 3 MM filter paper
- ❐ Nitrocellulose paper (0.45-μm pore size)
- ❐ Gloves
- ❐ Paper towels
- ❐ Razor blade

PROTOCOL

1. Photograph your gel (which should include molecular weight markers), making certain to include a ruler in the photograph to permit correct alignment of the bands on the blots with those on the gel. Carefully transfer your gel to a basin containing denaturation solution. Use enough solution to cover the gel completely. Soak for 30 minutes with occasional gentle agitation. Wear gloves when handling the gel. Use a glass, plastic, or cellophane wrap support when transferring the gel.

2. Place your gloved hand (or a support) over the gel and gently decant the solution. Rinse once with distilled water. After removing the water, soak the gel in neutralizing solution for 45 minutes.

3. While the gel is soaking, cut six sheets of Whatman 3 MM filter paper. These should be equal in size to the gel. Using a razor blade or a scalpel, cut two sheets of NC paper. These should also be the size of the gel. It is critical that you **wear clean gloves whenever handling nitrocellulose**, as oils from your hands will cause blotches on your filter or autoradiograph. Using a pen or pencil, label the NC paper and indicate "top" or "bottom" on each sheet. Be careful not to tear the NC paper with the tip of your pen or pencil. Finally, cut a 2-inch thick stack of paper towels, also equal in size to the gel.

4. Place half of the stack of paper towels on the bench top. Wet 3 sheets of 3 MM paper in the neutralizing solution and then stack

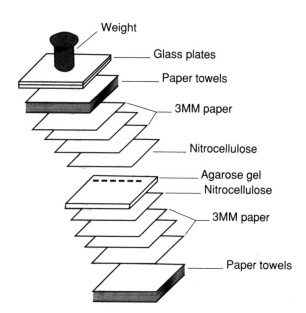

Weight
Glass plates
Paper towels
3MM paper
Nitrocellulose
Agarose gel
Nitrocellulose
3MM paper
Paper towels

Figure 8-1. Southern blotting set-up for bidirectional blotting of DNA from an agarose gel onto a nitrocellulose membrane.

them on top of the paper towels. Wearing gloves, transfer the "bottom" sheet of NC paper into the neutralizing solution. Once the paper is wetted, slip it under the gel. Transfer the NC paper and gel to the top of the wet filter papers.

5. Wet the "top" NC paper in neutralizing solution and place it on top of the gel. Smooth out any bubbles with your gloved hand, as bubbles will prevent transfer. Wet the remaining three sheets of 3 MM paper in neutralizing solution and place them on top of the NC paper. Put the remaining stack of paper towels on top. Two glass plates are balanced on top of this to ensure even contact. Put a weight on the plates. See the diagram of the Southern blotting set-up in Fig. 8-1.

6. Transfer is usually complete within 2 hours, but it can be allowed to proceed up to several days if necessary due to time restrictions.

DAY 2

Baking the Blot, Nick translation, and Biotinylation of DNA

A. Baking the Blot

MATERIALS

- ❏ 2x SSC (0.3 M NaCl, 0.03 M Na citrate)
- ❏ Gloves
- ❏ Filter paper
- ❏ Hand-held UV light

❑ Aluminum foil

❑ Ziploc bags

❑ Vacuum oven

PROTOCOL

1. Wearing gloves, remove the weights, paper towels, and filter papers surrounding the gel.

2. Use a pen to mark the location of the gel wells on the NC paper. This will allow you to orient your autoradiogram relative to the original gel. If ethidium bromide staining was not employed, this step will need to be done **prior** to separating the NC papers from the gel.

3. Remove the NC papers from either side of the gel and soak them for 5–10 minutes in 2x SSC.

4. Dry the blots on filter paper for 5 minutes. If the gel has previously been stained with ethidium bromide, visualize the DNA bands using a hand-held UV light.

5. Bake the blots on clean sheets of filter paper for 2 hours at 80°C under vacuum. Store the blots in sealed Ziploc bags at room temperature.

B. Nick Translation

Nick translation (Rigby *et al.*, 1977) is an efficient procedure for labeling double-stranded DNA with radiolabeled or biotinylated nucleotides. The procedure relies upon introducing single-stranded breaks (nicks) in the DNA backbone with *E. coli* DNase I and then replacing the nucleotide residues in front of the nick using labeled dNTPs and *E. coli* DNA polymerase I. Thus, the nick moves along the DNA strand ("translates") in a 5'→ 3' direction. Nick translation kits are available from a number of commercial suppliers, or the components can be prepared and stored at −20°C. If the latter approach is used as we do here, it is important that the concentration of the dNTPs be verified by optical density measurements. It is also critical that the activity of the DNase I be assayed to insure that it does not over- or underdigest the DNA sample. These procedures are detailed by Maniatis *et al.*, 1982. An alternative labeling approach, the random priming procedure, is becoming increasingly popular. It involves adding short oligonucleotide primers to the denatured DNA. The primers are elongated using the Klenow fragment of DNA polymerase I along with labeled dNTPs (Feinberg and Vogelstein, 1984).

MATERIALS

- ❏ TE (10 m*M* Tris–Cl, 1 m*M* EDTA, pH 8.0)

- ❏ Cloned DNA (stored in TE buffer at ~1 µg/µl)

- ❏ DNase I (1 mg/ml, stored at −20°C in 20 m*M* Tris–Cl, pH 7.6, 50 m*M* NaCl, 1 m*M* dithiothreitol, 100 µg/ml of BSA, 50% glycerol)

- ❏ Plexiglas shield

- ❏ 10× TM (5 *M* Tris–Cl, pH 7.8, 50 m*M* MgCl$_2$)

- ❏ 50 m*M* dithiothreitol

- ❏ 0.1 m*M* dATP

- ❏ 0.1 m*M* dGTP

- ❏ 0.1 m*M* TTP

Note: Adjust pH of dATP, dGTP, and TTP solutions to 7.0 using 0.05*M* Trisma base and pH paper.

123

❏ 5 mg/ml of acetylated bovine serum albumin (provided by most molecular biology supply companies)

❏ (α–^{32}P)dCTP (400 Ci/mmol, 10 mCi/ml)

❏ DNA polymerase I (6 units/μl)

❏ 15°C water bath

❏ 500 mM EDTA

❏ Bio-Rad P-30 resin equilibrated in TE

❏ Phenol saturated with 0.5 M Tris–HCl, pH 8

❏ Chloroform

❏ Microcentrifuge tubes (1.5 ml and 0.5 ml)

❏ Tabletop centrifuge

❏ 5 M ammonium acetate

❏ Sheared salmon sperm DNA at 100 μg/ml in TE buffer (DNA can be sheared by repeated passage through a syringe needle or by sonication)

❏ 100% ethanol

❏ 70% ethanol

❏ Scintillation counter

Note: DNA polymerase I, which contains both DNA polymerase and 5'→3' exonuclease activities, must be used instead of the large fragment (Klenow fragment) of DNA polymerase I, which does not contain the 5'→3' exonuclease activity.

Protocol

1. Nick translation of cloned DNA will be carried out as described below. These probes will be hybridized to the blots on Day 3. Add and mix each of the components to a microcentrifuge tube on ice.

CAUTION: Handle [^{32}P]-labeled compounds behind a Plexiglas shield. Labcoat, gloves, eye protection, and film badges should be employed.

Ingredients	Amount	Final Concentration
DNA (1 µg)	1 µl	
10x TM	5 µl	0.5 *M* Tris–Cl, pH 7.8, 5 m*M* MgCl$_2$
50 m*M* dithiothreitol	5 µl	5 m*M*
Distilled water	27.1 µl	
0.1 m*M* dATP	1.8 µl	3.6 µ*M*
0.1 m*M* dGTP	1.8 µl	3.6 µ*M*
0.1 m*M* TTP	1.8 µl	3.6 µ*M*
5 mg/ml of bovine serum albumin	0.5 µl	0.05 mg/ml
(α–^{32}P)dCTP (400 Ci/mmol, 10 mCi/ml)	4 µl	

Calculation: Using the specific activity of the radiolabeled nucleotide (Ci/mmol), the concentration of the undiluted nucleotide (mCi/ml), the molecular weight of dCTP (467 daltons), and the number of molecules in a mole (6.02 x 10^{23}), calculate the final concentration of the radiolabeled nucleotide in the nick translation reaction.

2. Dilute 0.8 µl of 1 mg/ml DNase I in 13.5 ml water. Add 1 µl of diluted DNase to the mixture.

3. Add 1 µl (6 units) of DNA polymerase I.

4. Incubate 70 minutes at 15°C.

5. Stop the reaction by adding 5 µl of 500 m*M* EDTA and 45 µl of water.

6. To remove unincorporated nucleotides, spin the nick translation mix over a 0.3 ml column of Bio-Rad P-30 resin equilibrated in TE. The P-30 is placed in a 0.5-ml microcentrifuge tube that has a small hole punctured in the bottom. This tube is slipped into a 1.5-ml microcentrifuge tube and centrifuged at 900 rpm for 3 minutes, in a table-top centrifuge, to remove the equilibration buffer. The column is then loaded with the sample and is centrifuged under the same conditions. Unincorporated nucleotides should be retained in the P-30 resin.

Note: If some of the resin is centrifuged into the 1.5 ml tube during removal of the equilibration buffer, discard the column and begin again. The needle hole was too large.

7. Extract the eluate with 1/2 volume of phenol and 1/2 volume of choloroform. Precipitate the upper aqueous layer by adding 0.4 vol of 5*M* ammonium acetate, 20 µg of carrier salmon sperm DNA, and 2.5 volumes of 100% ethanol (include the ammonium acetate volume in this final calculation). Precipitation is accomplished by incubation for 6 hours at –20°C or 30 minutes at –70°C.

Note: Ammonium acetate, unlike some other salts, does not cause unincorporated nucleotides to precipitate.

8. Pellet the DNA by centrifugation, wash once with 70% ethanol, dry, and resuspend in TE buffer. Count an aliquot of the DNA in the scintillation counter to determine the specific activity of the probe.

C. Biotinylation of DNA

Rather than use biotinylated NTPs to label the DNA by nick translation, this procedure uses photoactivatable biotin. The photoactivation reaction is simpler and quicker than nick translation.

MATERIALS

- ❐ Photoactivatable biotin (PAB) from Clontech Laboratories
- ❐ Cloned DNA (stored in TE at ~1 µg/µl)
- ❐ Microcentrifuge tubes
- ❐ Bucket of ice
- ❐ GE #RSM 275-watt sunlamp
- ❐ 0.1 *M* Tris–Cl, pH 9.0
- ❐ 2-Butanol
- ❐ 3 *M* sodium acetate, pH 5.6
- ❐ 100% ethanol

PROTOCOL

1. The PAB is light-sensitive and should be used under subdued light. Reconstitute the dry powder in water to 1 µg/µl. (This solution is stable at −20°C for up to 1 year.)

2. Mix 1 µl of DNA (1 µg) with 3 µg of PAB in a 1.5-ml microcentrifuge tube and place in an ice bucket, 10 cm below the sunlamp. Keep the cap of the tube open and irradiate for 15 minutes.

3. Add an equal volume of 0.1 M Tris–Cl (pH 9.0) and then add 100 µl of water.

4. Extract the unincorporated PAB with an equal volume of 2-butanol, retaining the lower, aqueous phase. Repeat once.

5. Precipitate the DNA with 0.1 volume of sodium acetate (pH 5.6) and 3 volumes of ethanol, followed by storage at −20°C for 2 hours or −70°C for 15 minutes.

6. Pellet the DNA in a microcentrifuge for 15 minutes. The biotinylated DNA pellet may appear pink. Dry *in vacuo* and resuspend in 500 µl of TE buffer. Store at −20°C.

Caution: Wear goggles to protect your eyes. It is probably safest to leave the room during the reaction, but it is essential that the ice bucket be completely filled with ice so that the heat lamp does not ignite the walls of the bucket.

DAY 3

Hybridization

Hybridization to nylon or nitrocellulose membranes containing nucleic acid is accomplished by adding single-stranded probe to the membranes which have been previously incubated with a prehybridization solution. Both prehybridization and hybridization solutions contain a buffer and various components designed to prevent adventitious binding of the probe to the filters. Formamide can be used to lower the temperature required for hybridization. Note that the phage plaque transfers (Lab IX) may be hybridized in the same containers as the Southern blots.

MATERIALS

❑ Prehybridization/hybridization solution [45% formamide, 5x SSPE [0.9 M NaCl, 50 mM H$_2$ PO$_4$, pH 7.4, 5 mM EDTA), 0.1% SDS, 5x Denhardt's solution (0.1% each of Ficoll, polyvinylpyrrolindone, and bovine serum albumin), and 100 μg/ml of denatured salmon sperm DNA]

Note: Hybrids containing biotinylated nucleotide residues are less stable than nonbiotinylated hybrids. To permit biotinylated DNA to hybridize efficiently, 45% formamide, rather than 50% formamide, is used in this solution. Fifty percent formamide may be used for a radioactive probe.

❑ Plastic bags that can be sealed using a heating device (e.g., Seal-n-Sav, Daizy Micro-Seal)

❑ 15 ml plastic tube

❑ Boiling water bath

❑ Bucket of ice

❑ Gloves

❑ Plexiglas shield

❑ Shaking incubator set at 42°C

PROTOCOL

1. Add 15 ml of prehybridization solution to each bag containing the blot. Remove bubbles and seal the bag using the heat-sealing device. Incubate at 42°C for at least 1 hour.

2. Remove prehybridization solution and replace with 10 ml of hybridization solution.

3. Pipet 1×10^6 counts per minute of radiolabeled probe or 200 ng of biotinylated DNA into a 15-ml plastic tube. Seal the tube with a plastic cap and poke a hole in the top with a syringe needle to prevent pressure build-up during boiling.

CAUTION: Handle radioactive material behind a Plexiglas screen. Wear gloves, glasses, lab coat, and film badge.

4. Denature the probe by placing the samples in the boiling water bath and heating for 10 minutes. Immediately transfer the tube to ice for 5 minutes (to prevent renaturation). Add 5 ml of hybridization buffer to the probe and transfer the sample to the plastic bag containing the filter. Reseal the bag.

5. Hybridize the blot for 1–5 days at 42°C. A shaking incubator is useful for distributing the probe evenly.

DAY 4

Washing the Blot

MATERIALS

- ❏ Glass baking dish (or plastic box)
- ❏ 0.1x SSC, 0.1% SDS (prewarmed to 50°C)
- ❏ 2x SSC, 0.1% SDS (room temperature)
- ❏ 2x SSC (room temperature)
- ❏ 0.15x SSC, 0.1% SDS (prewarmed to 50°C)
- ❏ Gloves
- ❏ Filter paper

❏ Cardboard

❏ Plastic wrap

❏ Tape

❏ X-Ray film folder

❏ X-Ray film (e.g., Kodak XAR5)

❏ Intensifying screen (e.g., Dupont Cronex Lightning Plus)

PROTOCOL

Blots probed with radioactive DNA

1. Remove each blot from its bag, tranfer to a glass baking dish, and rinse briefly in 50°C 0.1x SSC, 0.1% SDS. Remove this solution to radioactive waste and wash the blot three more times in the same solution. Agitate gently. Each wash is for 30 minutes. Carefully decant the waste while wearing gloves.

2. After the final wash, dry the blot on filter paper for 10 minutes. Carefully tape the filters to a cardboard backing. If available, a pen containing radioactive ink should be used to mark the cardboard in several areas to facilitate lining up the developed film with the filters. Cover with plastic wrap to prevent the filters from sticking to the X-ray film.

3. Place the cardboard containing the blots into an X-ray film folder. In the darkroom, place a piece of X-ray film over the filters. The film may be held in place with thumbtacks (the thumbtack holes will facilitate lining up the developed film with the filters). On top of this, place an intensifying screen. Close the folder and clamp it.

4. Store at −70°C. The intensifying screen reflects back radioactive disintegrations as light. Exposure at low temperatures prevents light scattering which would increase the length of exposure time needed.

Blots probed with biotinylated DNA

1. Wash filters in 2x SSC, 0.1% SDS for 3 minutes at room temperature (repeat once).

2. Wash filters in 0.15x SSC, 0.1% SDS for 15 minutes at 50°C (repeat once).

3. Store filters in 2x SSC at room temperature.

DAY 5

Developing the Film and Developing the Blot

In this procedure you will determine which DNA fragments hybridized to the probe you utilized. To find the molecular weights of the hybridizing bands, measure the distance each band has migrated from the marked loading well. Compare this distance to a standard curve of molecular weights vs. distance migrated, which was determined using the molecular weight markers on the stained gel.

MATERIALS

❏ Developer (e.g., Kodak GBX)

❏ Fixer (e.g., Kodak GBX)

❏ 3 developing trays

❏ Glass or plastic tray

❏ Gloves

❏ Buffer A (0.1 M NaCl, 0.1 M Tris–Cl, pH 7.5, 2 mM MgCl$_2$, 0.05% Triton X-100)

❏ Buffer B (buffer A with 3% bovine serum albumin)

❏ Buffer C (0.1 M NaCl, 0.1 M Tris–Cl, pH 9.5, 10 mM MgCl$_2$)

❏ SAP (streptavidin–alkaline phosphatase conjugate)

❏ NBT (nitro-blue tetrazolium), 50 mg/ml

❏ BCIP (5-bromo-4-chloro-3-inoyl phosphate) (SAP, NBT, and BCIP are available from Clontech Laboratories)

❏ 1 mM EDTA, pH 8

PROTOCOL

A. Developing the Film

1. Remove the folder from the freezer.

2. In a darkroom, remove film from folder and place in X-ray developer for 2–3 minutes.

3. Transfer the film to water for 30 seconds to wash off excess developer.

4. Fix in X-ray fixer for 3 minutes.

5. Wash in water for 2 minutes.

6. Air dry the film.

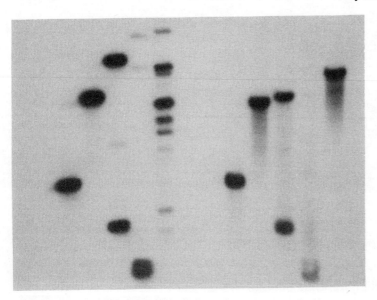

Figure 8-2. Autoradiograph of a Southern blot. Note that it is essential to locate where the wells on the gel were, relative to the bands on the autoradiograph. To determine which bands have hybridized, you must compare the distance migrated of the hybridizing bands to a photograph of the original gel (with a ruler present).

7. Place the film over the blot and mark locations of lanes. Compare location of bands to photographs of gels.

B. Developing the Blot

1. Block filter (to prevent nonspecific binding of the streptavidin conjugate) for 30 minutes in 20 ml of Buffer B, agitating occasionally. This may be performed in a glass or plastic tray. Alternatively Steps 1, 2, and 6 may be performed on top of a piece of plastic wrap. Place the side of the filter that faced away from the gel on top of the plastic wrap and spread the solutions directly on top of the blot.

2. Remove liquid and add SAP conjugate (10 ml of Buffer A containing 25 µl of conjugate). Incubate 25 minutes. Agitate occasionally.

3. Wash three times (10 minutes each) with 50–100 ml of Buffer A.

4. Wash once with 50–100 ml of Buffer C (5 minutes).

5. Dye solution is prepared by adding 64 μl of NBT and 32 μl of BCIP to 10 ml of Buffer C. Take care to prevent evaporation of these solutions; i.e., close lids quickly.

6. Incubate filter in dye solution under reduced light for 30 minutes to 3 hours.

7. Terminate the reaction by washing the blot with 1 mM EDTA. Store away from strong light after drying.

REFERENCES

A. P. Feinberg and B. Vogelstein, *Anal. Biochem.* **137**, 266 (1984).

T. Maniatis, E. F. Fritsch, and J. Sambrook, "Molecular Cloning—A Laboratory Manual." Cold Spring Harbor Lab., Cold Spring Harbor, New York, 1982.

P. W. J. Rigby, M. Dieckmann, C. Rhodes, and P. Berg, *J. Mol. Biol.* **113,** 237 (1977).

G. E. Smith and M. D. Summers, *Anal. Biochem.* **109,** 123 (1980).

E. M. Southern, *J. Mol. Biol.* **98,** 503 (1975).

LAMBDA PHAGE MANIPULATIONS

A. Phage Plating and Plaque Transfer

Constructing and screening a bacteriophage λ library is a common procedure utilized for isolating eukaryotic cDNAs and genes (Kaiser and Murray, 1985). A number of vectors are available that permit insertion of foreign DNA fragments between the terminal regions of the linear phage genome. Foreign DNA replaces λ sequences that are not essential for the lytic cycle. Typically, the intact phage DNA is cut with a restriction enzyme. The λ "arms" are then isolated away from the nonessential central region by sucrose gradient centrifugation or gel electrophoresis. The isolated arms (which may be phosphatase-treated to prevent ligation to each other, i.e., self-ligation) are added to genomic DNA or cDNA that possesses the same cohesive ends as the vector. Ligation of the foreign DNA into the phage arms is accomplished by the addition of T4 DNA ligase. The λ arms along with internal foreign DNA are then packaged using extracts from two *E. coli* strains that have each been infected with different λ mutants incapable of replicating or packaging their own DNA. The combined extracts complement each other and contain the phage components required to recognize the terminal *cos* sites (packaging signals) on the vector. Only DNA molecules with *cos* sites separated by ~50 kb are efficiently packaged. The size selection displayed by the extracts prevents packaging of phage arms that do not contain foreign

DNA, since such molecules are too small. Typically, inserts of 7–25 kb of foreign DNA can be cloned in these vectors. The recombinant DNA library is then titered by mixing with an *E. coli* culture, plating on agar, and incubating for several hours. Lambda arms and packaging mixes for library construction are commercially available from many molecular biological vendors.

Some λ vectors, such as λgt10 and λgt11, are chiefly used for cloning smaller DNA fragments, such as cDNA molecules (Huynh *et al.*, 1985). These vectors do not contain internal segments that are removed prior to library construction. Even though the digested arms of these λ vectors are treated with phosphatase, a background level of nonrecombinants is invariably obtained, either due to incomplete restriction enzyme digestion or incomplete phosphatase treatment. However, some vector/host systems permit direct selection against nonrecombinants. In λgt10, for instance, foreign DNA is cloned into the λ *cI* gene. Phage that do not have their *cI* gene disrupted are incapable of forming plaques on *hfl* (high frequency of lysogeny) bacteria and can be eliminated from consideration. For λgt11, foreign DNA is cloned into the *lac Z* gene. One can determine the percentage of recombinant phage in a library by plating on IPTG (to induce the *lac* promoter) in the presence of the chromogenic substrate X-gal. Plaques that are blue under these conditions do not have foreign DNA disrupting the *lacZ* gene.

Once a λ library is constructed, it can be screened with a hybridization probe or, if it is an expression library, with an antibody. Expression libraries permit the transcription and translation of inserted DNA. Lambda expression vectors (e.g., λgt11) contain the *lacZ* gene upstream of a unique restriction site. By inserting foreign DNA into the restriction site, a fusion gene can result. Transcription can be induced by the addition of IPTG. A fusion protein, with the same amino acid sequence as the cloned foreign DNA, will result, on average, one in six times, when both the correct 5'→3' orientation and the correct reading frame are obtained. When screening an expression library, phage plaques are allowed to form prior to induction of the *lacZ* promoter in order to insure that clones containing potentially lethal fusion proteins are able to develop. Expression libraries are usually plated on protease-deficient (*lon⁻*) strains of *E. coli* in order to reduce degradation of the fusion protein. Induction of fusion protein synthesis is typically accomplished by placing IPTG-impregnated nitrocellulose or nylon filters on top of the phage plate for several hours. Proteins are then fixed on the filters, the filters are "blocked" with a protein solution to prevent nonspecific antibody retention, and an antibody is added. These "Western" blots (solid matrices with proteins fixed to them) are washed and reacted with a second antibody which recognizes the first antibody. The second antibody contains an enzymatic or radioactive tag so that locations where it binds can be determined by a colorimetric assay or autoradiography (see Huynh *et al.* [1985] for details of these procedures).

Hybridization screening of a λ phage library involves hybridizing a radiolabeled or biotinylated probe to filters containing phage DNA that has been denatured and fixed *in situ*. The hybridization, washing, and detection procedures are those used in the Southern blotting experiment.

Once a putative positive clone is detected, the plaque must be purified to homogeneity. Typically, many thousands of plaques are screened on each plate, and it is impossible to precisely localize the positive signal by lining up the stained filter or exposed X-ray film with the original phage plate. Therefore, a plug of phage plaques (typically ~5 mm in diameter) is removed from the approximate location of the positive signal. Phage are eluted from the plug and the lysate is titered. A plate with a few hundred plaques is prepared and rescreened in order to identify a single positive plaque.

In this experiment, you will dilute and plate a phage library in order to screen it by hybridization. Note that top agarose is used in this procedure (rather than agar) since it tears less easily and does not stick as readily to nitrocellulose membranes.

Preparation of Cells

MATERIALS

❏ Appropriate cell line (in this case, *E. coli C600 F⁻, thi-1, thr-1, leuB6, lacY1, tonA21, supE44, λ⁻*)

❏ LB agar plates (15 g/l agar)

PROTOCOL

1. Streak out C600 on an LB plate. Store at 37°C overnight.

Overnight Culture

MATERIALS

❏ LB medium

❏ Sterile10-ml tube

❏ 37°C shaker bath

PROTOCOL

1. Transfer a single colony into 5 ml of LB medium.

2. Shake overnight at 37°C.

Plating the Phage

MATERIALS

❑ NZCYM medium [per liter: 10 g of NZ amine (Type A hydroly-sate of casein), 5 g of NaCl, 5 g of yeast extract, 1 g of casamino acids, 2 g of $MgSO_4 \cdot 7H_2O$, adjusted to pH 7.5 with NaOH]

❑ 37°C shaker bath

❑ Spectrophotometer

❑ 100-ml graduated cylinder (sterile)

❑ Sterile Oakridge tubes

❑ Sorvall superspeed centrifuge, cooled to 4°C

❑ Sorvall SS-34 rotor, cooled to 4°C

❑ Sterile SM buffer, ice cold (100 mM NaCl, 10 mM $MgSO_4$, 50 mM Tris–HCl, pH 7.5, 0.01% gelatin)

❑ 0.1- and 1-ml pipets (sterile)

❑ Sterile 1.5-ml microcentrifuge tubes

❏ 4 NZCYM/agar or agarose plates (15 g of agar or agarose/liter; plates prewarmed to 37°C)

❏ 4 vials of sterile top agarose/NZCYM (7.5 g of agarose/liter. Use 3 ml for each 10-cm plate). Liquefy by boiling. Cool by incubating at 45°C.

❏ Phage library of your choice at 10^5 plaque-forming units/ml.

PROTOCOL

1. Dilute 0.8 ml of overnight culture into 200 ml of NZCYM in a 1-liter flask.

2. Shake at 37°C until $OD_{600} = 0.25$. This should take ~3 hours.

3. Transfer 25 ml of the culture into a sterile Oakridge tube.

4. Centrifuge at 5000 rpm (3000 g) for 10 minutes at 4°C in an SS-34 rotor. Remove the supernatant.

5. Resuspend in cold SM buffer to 1.5×10^9 cells/ml. (Assume that an OD_{600} of 0.25 equals 2×10^8 cells/ml.)

6. Distribute 0.1 ml of cells into four sterile 1.5-ml microcentrifuge tubes.

7. Dilute an aliquot of the phage library with SM buffer to 10^4 and 10^3 pfu/ml.

8. Add: 0.1 ml of SM to bacterial tubes 1 and 2, 0.1 ml of 10^4 pfu/ml to bacterial tube 3, 0.1 ml of 10^3 pfu/ml to bacterial tube 4.

9. Mix and incubate for 20 minutes at 37°C to allow the phage to adsorb to the bacteria.

10. In the meantime, label four prewarmed NZCYM plates (one for each tube).

11. Quickly add the contents of one microcentrifuge tube to a vial of soft agarose (45°C). Mix and immediately pour onto the prewarmed NZCYM plate. With the plate on a flat surface, swirl gently to ensure an even distribution of agarose. Tubes 1 and 2 serve as a control so that you can master the technique of pouring the top agarose quickly and evenly.

Note: Note that an even surface is essential for transferring phage particles to filters (making plaque lifts) as described below. Agarose is the preferred medium for the phage plating material, since it gives less background hybridization than agar.

12. Repeat Step 11 for tubes/plates 2–4.

13. After cooling at room temperature (5–10 minutes), the plates can be inverted and stored at 37°C.

Note: After an overnight growth period, refrigerate the phage plates at 4°C to stop growth and to harden top agarose.

DAY 2

Plaque Transfer

MATERIALS

❐ Gloves

❐ Nitrocellulose circular filters (a size that will cover the entire surface of the plate used)

❐ Sterile needle

❐ Petri plates

❐ 0.1 *N* NaOH, 1.5 *M* NaCl

❐ 0.5 *M* Tris–Cl, pH 7.4, 1.5 *M* NaCl

- ❏ 2x SSC (0.3 *M* NaCl, 0.03 *M* Na citrate)
- ❏ Filter paper
- ❏ Aluminum foil
- ❏ Ziploc plastic bag

PROTOCOL

1. Pick a plate with about 100 plaques for transferring phage to filters. (For screening a library, you would usually use a plate with about 10,000 plaques.)

2. **Wearing gloves**, take a circular nitrocellulose (NC) filter and label it with your initials.

3. Carefully place the NC filter on top of the plate. Do this slowly so that no air bubbles form. It should take a couple of minutes.

4. Label and place a second NC filter on top of the first.

5. Mark the orientation of filters on the plate by stabbing three holes through the filter and the agar using a sterile needle. Do not make the holes symmetrical or it will be difficult to determine the correct orientation. Mark the bottom of the plate to indicate the location of the holes.

6. Return the plates to the refrigerator and let them adsorb for at least 20 minutes.

7. Carefully remove the top filter and float face-up in a petri plate containing 0.1 *N* NaOH, 1.5 *M* NaCl. (This serves to denature the phage DNA.) After a few seconds, sink the filter. Soak for 30 seconds.

8. Transfer to 0.5 *M* Tris–Cl, pH 7.4, 1.5 *M* NaCl for neutralization. Soak for 30 seconds.

9. Remove to 2× SSC. Soak for 30 seconds.

10. Air dry on filter paper.

11. Repeat Steps 7–10 with the bottom filter. Be careful not to lift the top agarose. If this occurs, remove it from the filter during the 2× SSC wash using your gloved hand.

12. Sandwich both NC filters between clean filter paper. Wrap loosely in aluminum foil (so an exit is provided for escape of water). Bake at 80°C *in vacuo* for 2 hours.

13. Store in a sealed plastic bag at room temperature until ready to hybridize with probe.

14. Hybridization and signal detection should be carried out as described in the Southern blotting protocol (Lab VIII, Days 3–5).

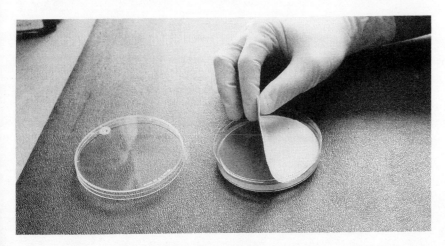

Figure 9-1. Placing a nitrocellulose filter onto an agar plate to perform a plaque lift.

B. Bacteriophage Lambda Miniprep

Once a positive signal is detected in a phage library and the plaque of interest is purified, it is usually appropriate to purify the phage and analyze the inserted foreign DNA by restriction enzyme mapping. Large-scale preparations involve growing the host in liquid culture along with phage so that a lysate is formed. The phage particles are isolated on a cesium chloride gradient and the DNA is extracted by protease, phenol, and chloroform treatments (Miller, 1987). Satisfactory small-scale preparations can be made that permit rapid analysis of a number of clones of interest. In this procedure, you will perform a bacteriophage λ miniprep from a λ lysate contained on one agar plate. You will isolate λ DNA from the bacteriophage and analyze the phage DNA by restriction enzyme digestion and gel electrophoresis.

DAY 1

Plating the Phage

MATERIALS

- ❏ Appropriate bacterial cell line (in this case, C600)
- ❏ Top agarose in small vials (7.5 g/l agarose in NZCYM medium, use 3 ml per 10-cm plate)
- ❏ NZCYM-containing agarose plates (prewarmed to 37°C)
- ❏ Desired phage isolate at a known titer
- ❏ 37°C incubator

PROTOCOL

1. Infect C600 with phage and plate on an *agarose*-NZCYM plate as described in the phage plating and plaque transfer protocol. Agar often contains inhibitors of restriction enzymes, so agarose is utilized in this procedure. Plate 5000 plaque-forming units pfu on one plate and 20,000 pfu on a second plate.

2. Incubate at 37°C overnight.

DAY 2

Making a Plate Lysate

MATERIALS

☐ Sterile SM buffer (100 mM NaCl, 10 mM MgSO$_4$, 50 mM Tris–HCl, pH 7.5, 0.01% gelatin)

☐ Oakridge tubes (sterile)

☐ Sterile 5 ml pipets

PROTOCOL

1. Transfer 5 ml of SM buffer onto one plate. Use the one with the lower number of phage if it is completely lysed. Shake gently every 10 minutes or so for 2 hours.

2. Transfer the lysate to a sterile Oakridge tube.

3. Refrigerate at 4°C until you are ready to begin phage isolation or proceed immediately to Day 3.

Phage Isolation

MATERIALS

- ❑ Sorvall superspeed centrifuge at 4°C
- ❑ Sorvall SS-34 rotor (cooled to 4°C)
- ❑ Oakridge tubes (sterile)
- ❑ RNase A + DNase I solution (100 µg/ml each)
- ❑ 20% (w/v) polyethylene glycol, 2 *M* NaCl in SM buffer
- ❑ 1.5-ml microcentrifuge tubes (sterile)
- ❑ 37°C water bath
- ❑ Microcentrifuge

PROTOCOL

1. To remove bacterial debris, centrifuge the refrigerated tube from Day 2 at 8000 rpm (7700 *g*) for 10 minutes in the SS-34 rotor, at 4°C.

2. Transfer the supernatant to a clean Oakridge tube and add 50 µl of 100 µg/ml RNase A + DNase I. Mix and incubate at 37°C for 30 minutes. (This step degrades *E. coli* DNA and RNA, but does not affect the λ DNA, which is still within the protein coat of the phage particles.)

3. Add 5 ml of 20% polyethylene glycol, 2 *M* NaCl in SM buffer.

Mix and store on ice for at least 1 hour to precipitate the phage.

4. Centrifuge at 10,000 rpm (12,000 *g*) in an SS-34 rotor at 4°C for 20 minutes.

5. Carefully remove the supernatant as completely as possible, and discard.

6. Resuspend the pellet in 0.5 ml of SM buffer and transfer it to a microcentrifuge tube.

7. Refrigerate at 4°C until you are ready to proceed with phage DNA isolation.

DAY 4

Phage DNA Isolation

MATERIALS

- ❏ 10% (w/v) SDS (sodium dodecyl sulfate)
- ❏ 0.5 *M* EDTA, pH 8.0
- ❏ Gloves
- ❏ Phenol saturated with 0.1 *M* Tris–Cl, pH 7.5
- ❏ Phenol:chloroform (1:1) saturated with 0.1 *M* Tris–Cl, pH 7.5
- ❏ Chloroform
- ❏ Isopropanol
- ❏ 70% ethanol
- ❏ TE buffer (10 m*M* Tris–HCl, 1 m*M* EDTA, pH 8.0)

❏ Microcentrifuge at room temperature

❏ Microcentrifuge at 4° C

❏ 68°C water bath

❏ Dry ice–ethanol bath

❏ Vacuum drying apparatus

PROTOCOL

1. Add 5 µl of 10% SDS and 5 µl of 0.5 *M* EDTA to your sample. Incubate at 68°C for 15 minutes to lyse the phage particles.

2. **Wear gloves, lab coat, and safety glasses.** Extract protein contaminants with phenol **(work in fume hood)** by adding 2 volumes of phenol. Vortex well and microcentrifuge for 2 minutes. Remove the aqueous upper layer to a clean microfuge tube.

3. Repeat Step 2 using a 1:1 mixture of phenol and chloroform. Repeat again.

4. Repeat Step 2 using 2 volumes of chloroform.

5. Add 1 volume of isopropanol. Place in the dry ice–ethanol bath for 20 minutes to precipitate the DNA.

6. Centrifuge in the microcentrifuge at 4°C for 15 minutes.

7. Remove the supernatant and add 70% ethanol. Centrifuge for 5 minutes.

8. Remove the supernatant and dry the pellet *in vacuo*.

9. Resuspend the pellet in 50 µl of TE buffer. Use 10 µl for each restriction digest.

10. Treat the DNA with a restriction enzyme that will allow identification of recombinant phage when the digestion products are separated via electrophoresis through an agarose gel. To prevent degradation, restriction digests should be allowed to proceed for only 2 hours. Store digested or undigested DNA at $-20°C$.

DAY 5

Running Gel, Staining, and Photography

Load the digested and undigested DNA on a 12-cm agarose gel along with size markers. It would be useful to include a lane of gradient-purified DNA that has been digested with the same enzyme. Due to the cohesive ends of the λ DNA, the phage arms may appear as a very large fragment. To prevent this, DNA should be heated for 5 minutes at 65°C just prior to gel loading. Perform electrophoresis as previously described (Lab 2, Days 4 and 5).

REFERENCES

T.V. Huynh, R. A. Young, and R. W. Davis, *in* "DNA Cloning: A Practical Approach" (D. M. Glover, ed.), Vol. 1, pp. 49–78. IRL Press, Washington, D.C., 1985.

K. Kaiser and N. E. Murray, *In* "DNA Cloning: A Practical Approach" (D. M. Glover, ed.), Vol. 1, pp. 1–47. IRL Press, Washington, D.C., 1985.

H. Miller, *Methods Enzymol.* **152,** 145 (1987).

USE OF COMPUTERS IN A MOLECULAR BIOLOGY LABORATORY

Douglas W. Smith

Department of Biology and Center for Molecular Genetics
University of California, San Diego
La Jolla, CA 92037

With the advent of recombinant DNA and DNA sequencing techniques, coupled with the concomitant exponential increase in DNA and protein sequence information, the use of computers in a DNA- and protein-oriented molecular biology laboratory has become nearly indispensible. Tasks routinely performed by computers in DNA and protein molecular biology work can be categorized as follows: (1) planning strategies for projects and parts of projects, for example, determining a strategy for cloning a particular DNA fragment; (2) manipulation and analysis of raw experimental data; (3) analysis of a DNA or RNA sequence, including searches for sites, translation into protein sequences, and searches for direct and inverted repeats; (4) analysis of a protein sequence, including tertiary structure as well as secondary structure; (5) comparison of new DNA, RNA, or protein sequences with those present in libraries, with determination of evolutionary relationships; (6) analysis of molecular interactions, for example, of a protein with its substrates or of a DNA fragment with a DNA binding protein; and (7) preparation of manuscripts, posters, and other documents, including both data figures and complex cartoon figures of DNA constructions, etc., for dissemination of research results.

Use of computers involves three operational phases: (1) input of data, (2) execution of programs, and (3) output of transformed data from the computer. For DNA and protein work, input comes mainly from three sources: (1) a device inputs information into the computer directly from experimental data, for example, input of DNA sequence information from a sequencing gel using a digitizer device; (2) the user types the information into the computer, either directly into a program or into a file which is then used by the program; and (3) information is obtained by a program from files already in the computer, either created by the user or created from databases or other sources. When the input information is DNA and protein sequence information, different programs often require different "formats" for a given sequence.

Programs are either written by the user or obtained from elsewhere, either purchased from vendors or obtained from other research workers and groups. These vary from performance of single tasks to large, total "environment" programs, in which the user usually performs all operations from a series of menus with no direct interaction with the operating system of the computer. The latter often uses many programs, the use of which, due to the menu-based environment, is invisible to the user.

Output from the program(s) is either to the terminal screen, to an output file, or directly to a printer, plotter, or other peripheral device. The format for presentation of the output information is dictated by the program; many programs provide format options for the user. Output is either alphanumeric (words, letters, and numbers) or graphical. A few computer systems, including the Apple Macintosh and the Sun Workstations, are oriented around a graphical approach; most, however, are predominantly alphanumeric. Unlike the alphanumeric type of data, with the ASCII (American Standard Code for Information Interchange) standard, no general standards exist for graphical output. Hence, such output is often both computer- and software-specific, as, for example, with the Macintosh interface. Familiarity with these basic ideas is of great value in both choosing an appropriate computer system with associated software for the user's specific needs and for understanding why there is such a plethora of options and systems and why these are so often incompatible with each other.

Computers vary in size and capability, from the microcomputers (such as the Apple Macintosh and IBM PC systems) through the minicomputers (such as the Digitial Equipment Corporation VAX computers) to the mainframes (such as the IBM systems) and the supercomputers (such as the Cray2). Everyone is aware of the incredible rate at which computer system development is progressing. Recent Apple Macintosh computers (the SE and Macintosh II) and "386" computers such as the Campaq Deskpro 386 perform most tasks two to ten times faster than a minicomputer such as a VAX 11/780. Hard disk and optical disk technology similarly now permits sufficient storage capability to house current DNA and protein sequence databases. However, microcomputers are largely "single-user, single-task" machines, although this too is changing. Minicomputers and larger computers are multiple-user (usually via "time sharing"), multiple-tasking machines. Thus, a single molecular biology system can be simultaneously accessed and used by typically between 5 and 30 users. Most users of minicomputers use a central facility with concomitant "recharge" costs; after the initial purchase, a microcomputer system for a laboratory is largely "free" of additional costs.

Excellent programs, both complete sets and individual programs, exist for DNA molecular biology for both microcomputers and minicomputers, and new programs are being written continously. In general, the tasks involved do not require the resources of mainframe computers or of supercomputers. This may change. In particular, tasks which require multiple passes through large sequence databases, even with their current sizes, can be prohibitive both in time and cost. With the "human genome sequencing initiative" and sequencing of many large genomes, the sizes of current sequence databases will increase dramatically. Even at the current time, the databases are doubling in size about every 1.5 years. Rapid and efficient creation of up-to-date national and international databases, with rapid and efficient access to needed information within these databases, will require new approaches to both software (programs) and hardware (the computers and their physical peripheral devices) needs. Parallel processing, now currently done mainly within a supercomputer environment, is one approach with much promise, and indicates one potential important use of supercomputers in this type of molecular biology.

A second class of tasks which currently requires specific types of computers and software is that of analyzing 3-dimensional structure of proteins and their interactions with other molecules. Usually the user desires to vary parameters such as coordinates of specific protein residues or of substrate molecules and visually see on the screen the effects on the 3-dimensional structure. This requires graphics capability of high resolution, usually in color with appropriate shading for 3-D effects, as well as immediate response of the image to input changes from the user. Until very recently, microcomputers have had neither the resolution nor the speed necessary for such tasks, and appropriate systems have costs equivalent to those of minicomputers, although they are largely single-user instruments. This, however, is rapidly changing; a general principle is that costs of a given computer capability are decreasing 10-fold every 10 years. Thus, the required graphics and real-time capability may soon be available on microcomputers, with costs in the range of $5,000–$10,000.

Both commercially available and public domain programs are available for microcomputers and minicomputers. For the MS-DOS operating system microcomputers (such as the PC AT and Compaq Deskpro 386), six sets of commerical programs are most popular. These include the MicroGenie programs from Beckman, the PC/Gene programs from IntelliGenetics, the DNASIS programs from Hitachi or Pharmacia/LKB, the IBI/Pustell programs from International Biotechnologies, the DNA-Star programs from DNA-STAR (Blattner and Schroeder, 1984), and the Staden programs, somewhat modified (Staden Plus), from Bio-Rad. Each of these provides a rather complete environment for most usual tasks involving DNA and protein sequences, including digitizer entry and database handling of sequences from a DNA sequencing project, searching for homologies in the GenBank DNA (Bilofsky and Burks, 1988) and PIR protein (Sidman *et al.*, 1988) databases made available with these programs, and in some cases programs for drawing plasmid maps, etc. Considerable use is made of the graphics capabilities of these microcomputers. The programs are menu-driven and in general very user-friendly. They suffer from the limited memory capability of MS-DOS, and library search programs tend to run slowly. Also, the user is dependent on the vendor for updates of both programs and databases; there usually is a lag time of several months behind a direct subscription to the databases. Graphics output also is slow. The output formats vary considerably between these sets of programs, as do the algorithms used by some of the programs that perform similar tasks. The Micro-Genie programs were largely developed from the Korn–Queen programs (Queen and Korn, 1980), one of the first sets available for an IBM mainframe environment. The IBI/Pustell programs and the Bio-Rad Staden programs were developed, respectively from the sets of programs developed by Pustell (Pustell and Kafatos, 1984) and Staden (Staden, 1984a,b) for the minicomputer environment. DNASTAR provides hardware as well as software, but the programs can be purchased by themselves.

For the Apple Macintosh, two sets of commerical programs, DNA Inspector II and MacGene, have been the most popular. Both perform nearly all memory-nonintensive tasks very well, with a strong graphics orientation and a rather typical mouse-oriented Macintosh interface. However, they have very limited capability for database searching and for data entry for DNA sequencing projects. Two additional sets of programs have recently become available, Clone and MacMolly. Clone is oriented around plasmid analysis and drawing of plasmid maps, but has some limited protein sequence analysis capability. MacMolly is a more complete system of programs, rivaling the ones for the MS-DOS microcomputers.

The most popular commercial set of programs for minicomputers is the University of Wisconsin Genetic Computer Group (GCG) set of programs for VAX/VMS computers (Devereux *et al.* 1984). A user-friendly menu system has been developed, largely isolating the user from the VMS operating system. A built-in editor permits entry and editing of sequences. A plasmid map facility, and considerable graphics capability, in general, is included. Graphics output, however, is dependent on specific plotters and printers. The standard databases are included for database searching, and are updated biannually to subscribers. Capability for database handling of DNA sequencing projects has recently been added. Although less popular in university settings because of the expense, the most complete sets of programs are those from IntelliGenetics and from Battelle. These will basically "do everything," including extensive high-resolution graphics, and extensive vendor support is provided. A large subset of the IntelliGenetics programs is available via BioNet (Smith *et al.* 1986), an NIH-supported communication network to IntelliGenetics computers. BioNet suffers from relatively few access ports and relatively little vendor support.

Many sets of programs exist in the public domain for both micromputers and minicomputers, and in several cases, the commercially available programs are developments from the public domain programs. Most of these are described in molecular biology journals such as *Nucleic Acids Research* or in computer journals with an orientation toward molecular biology, such as *Computer Applications in the Biosciences* (*CABIOS*). The January issues of *Nucleic Acids Research* for 1982, 1984, 1986, and 1988 are devoted completely to computer applications in molecular biology. The user, thus, has a very large set of programs to sample, and this set is ever expanding. These noncommercial programs suffer mainly from nonprofessional documentation and lack of extensive support, maintenance, and updating of the programs. Of the many sets of relatively comprehensive programs, the Mount–Conrad (Mount and Conrad, 1986) programs from the University of Arizona and the Stephens (Stephens, 1985) programs from the NCI–Frederick Cancer Research Facility have been most useful for the MS-DOS environment. The Los Alamos SEQ alignment programs for VAX/VMS computers have also now been expanded into a large set of programs, with inclusion of several sets of DNA and protein databases (Kanehisa *et al.* 1984).

A general consideration regarding the use of commercial programs is that the user is "locked in" to the choices of the vendor. This includes input formats and capabilities, choice of algorithms and programs in general, and output formats. Often it is difficult to take output from one commercial set of programs and use this for input into other programs from a different source. Usually the options provided for program parameters are not as extensive as desired. The user is dependent on the vendor for program and database updates, which are usually less often than desired. The vendors are seldom willing to provide program code for user modification.

Alternative possibilities include use of home-written programs, use of public domain programs to supplement commerical program sets, or use of more than one commercial program set. An approach taken at the University of California at San Diego was to obtain several sets of well-proven, powerful public domain programs and to integrate these into a common system under the VAX/VMS environment (Smith, 1988). Programs chosen include some redundancy for a given task, to permit the user comparison of different algorithms and choice of output formats. System integration and ease of use are solved by a combination of VAX/VMS features, including use of command language files, help library, utility programs, and program and system documentation. The system runs via a modified menu system, where users access documentation either via the help facility or via use of the VMS command TYPE. Similarly, programs are run via use of the VMS command RUN. Programs chosen which have graphics capability nearly always use the Tektronix 4014 graphics standards, permitting output from any such program to common plotters or graphics printers. The sets of current programs include the Staden programs (Staden, 1980, 1984a,b), the Los Alamos SEQ homology programs (Kanehisa *et al.* 1984), the Pearson MAP programs (Pearson, 1982), the PHYLIP phylogeny programs (Felsenstein, 1982), the Doolittle NEWAT homology programs (Doolittle, 1981; Feng and Doolittle, 1987), the PIR protein alignment programs (George *et al.* 1986), the Pearson-Lipman FAST search programs (Lipman and Pearson, 1985; Pearson and Lipman, 1988), the Ooi and Takanami secondary structure program SECSTRUC (Ooi and Takanami, 1981), the Zuker RNA folding programs (Zuker and Stiegler, 1981), and the Finer–Moore and Stroud (1984) AMPHIPATHICITY programs, as well as some locally written programs. Databases and libraries include the PIR protein libraries, the GenBank DNA and RNA libraries, and the Doolittle NEWAT protein library. A utility program also translates the GenBank DNA libraries, providing a GenBank protein library. Addition of new sets of programs, removal of old programs, and database updating are all relatively easily accomplished. No rewriting of the program code is needed. The entire system is easily ported to other VAX/VMS computers, requiring, at best, change of only one line in one command file. This approach has provided the flexibility needed to obtain a single system which is responsive to the needs of a variety of types of molecular biologists. The flexibility has some cost in loss of user-friendliness in the VAX interface: users must learn to use a few verbs such as TYPE, RUN, and PRINT.

Use of this system also exemplifies some of the possibilities for combined use of different computer systems. For example, users can obtain output from a graphics program on the screen of a Macintosh used as a terminal to access the VAX. This screen image can then be obtained as a Macintosh file via an appropriate "screen dump"; the Macintosh file can then be used as input for a Macintosh graphics program such as MacDraft or MacDraw, with a final result sent to a LaserWriter printer. The result is a publishable figure. An example of such an approach is shown in Fig. A-1. This figure was generated from the VAX Staden program ANALYSEQ (Staden, 1984b) in the above described DNA system, followed by screendump to a Macintosh Plus and further modification with SuperPaint. The figure clearly shows that the *E. coli* origin of DNA replication, *oriC*, does not encode any polypeptides. Similarly, another user might prefer to use a Macintosh commercial program for routine operations such as searching for restriction sites, because of a preferred output format, but would then in turn use the Staden DBSYSTEM on the VAX for database handling of a DNA sequencing project. Many such combinations of program and computer systems and software are possible, and often desirable.

Thus, a wide range of computers and associated peripheral devices and software is available to satisfy the needs of the molecular biologist working with DNA

A. *E. coli* origin region, 5435 bp, sequence reversed from the *E. coli* map.

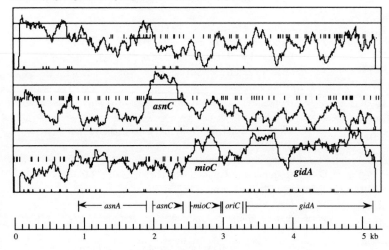

B. *E. coli* origin region, 5435 bp, direction as in the *E. coli* map.

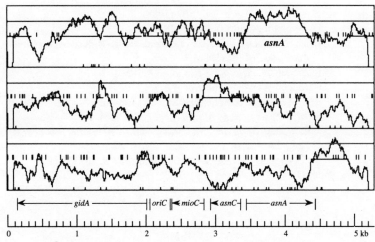

Figure A-1. Gene predictions for the nucleotide sequence (Bühk and Messer, 1983; Walker *et al.* 1984) in the *Escherichia coli* DNA replication origin region using the positional base preference method of Staden (1984b). This method is one of the options in ANALYSEQ (Staden, 1984a), a general sequence analysis program. The *x* axis represents the DNA sequence and the probability of coding is plotted in the *y* direction. This method assumes that there is a typical amino acid composition and no codon preference and takes advantage of the uneven use of amino acids by proteins. Probabilities of coding are calculated by sliding a window of 67–99 codons along the sequence and scoring each of the three reading frames at each window position. Probabilities are plotted for each reading frame, the three in one direction in A, the three in the opposite direction in B, one above the other. Potential initiation codons ATG and GTG are marked as vertical bars along the base of each plot. The point showing which of the three frames in A and B scored highest is plotted at a height corresponding to the expected value for a random noncoding DNA sequence, and stop codons are marked along this line as vertical bars. A third horizontal line corresponds to the expected mean value for coding (52.6%). The known genetic regions, with direction of gene expression, are indicated below the plots for each strand.

and protein sequences. Solutions can be found which combine computer use from the beginning of data entry to the final steps of manuscript preparation. Adequate to excellent commerical sets of programs exist to satisfy a wide range of budgets, as well as an even wider range of public domain or very inexpensive programs and sequence libraries. Often, combinations of microcomputers and minicomputers, with the microcomputers being used as terminals to interact with the minicomputers, provide optimal solutions to the needs of users. As microcomputers become even more powerful, multiuser and multitasking environments using only microcomputers will be formed. These are likely to include solutions for users needing high-resolution graphics displays of macromolecular 3-dimensional structure, visualized in real time. Solutions to access and efficient use of the increasingly large DNA and protein sequence databases, and to the protein 3-dimensional coordinate databases, will take longer, requiring new approaches. Tasks requiring extensive use of such databases will possibly be the last to migrate to the smaller machines.

REFERENCES

H. S. Bilofsky and C. Burks, *Nucleic Acids Res.* **16**, 1861 (1988).

F. R. Blattner and J. L. Schroeder, *Nucleic Acids Res.* **12**, 615 (1984).

H. -J. Bühk and W. Messer, *Gene* **24**, 265 (1983).

J. Devereux, P. Haeberli, and O. Smithies, *Nucleic Acids Res.* **12**, 387 (1984).

R. F. Doolittle, *Science* **214**, 149 (1981).

J. Felsenstein, *Q. Rev. Biol.* **57**, 379 (1982).

D. –F. Feng and R. F. Doolittle, *J. Mol. Evol.* **25**, 351 (1987).

J. Finer-Moore and R. M. Stroud, *Proc. Natl. Acad. Sci. U.S.A.* **81**, 155 (1984).

D. G. George, W. C. Barker, and L. T. Hunt, *Nucleic Acids Res.* **14**, 11 (1986).

M. Kanehisa, P. Kelin, P. Greif, and C. DeLisi, *Nucleic Acids Res.* **12**, 417 (1984).

D. J. Lipman and W. R. Pearson, *Science* **227**, 1435 (1985).

D. W. Mount and B. Conrad, *Nucleic Acids Res.* **14**, 443 (1986).

T. Ooi and M. Takanami, *Biochim. Biophys. Acta* **655**, 221 (1981).

W. R. Pearson, *Nucleic Acids Res.* **10**, 217 (1982).

W. R. Pearson and D. J. Lipman, *Proc. Natl. Acad. Sci. U.S.A.* **85**, 2444 (1988).

J. Pustell and F. C. Kafatos, *Nucleic Acids Res.* **12**, 643 (1984).

C. L. Queen and L. J. Korn, *Methods Enzymol.* **65**, 595 (1980).

K. E. Sidman, D. G. George, W. C. Barker, and L. T. Hunt, *Nucleic Acids Res.* **16**, 1869 (1988).

D. W. Smith, *CABIOS* **4**, 212 (1988).

D. H. Smith, D. Brutlag, P. Friedland, and L. H. Kedes, *Nucleic Acids Res.* **14**, 17 (1986).

R. Staden, *Nucleic Acids Res.* **8**, 817 (1980).

R. Staden, *Nucleic Acids Res.* **12**, 521 (1984a).

R. Staden, *Nucleic Acids Res.* **12**, 551 (1984b).

R. M. Stephens, *Gene Anal. Tech.* **2**, 67 (1985).

J. E. Walker, N. J. Gay, M. Saraste, and A. N. Eberle, *Biochem. J.* **224**, 799 (1984).

M. Zuker and P. Stiegler, *Nucleic Acids Res.* **9**, 133 (1981).

SAFETY IN THE RECOMBINANT DNA LABORATORY

Workers in recombinant DNA laboratories must be cognizant of safe handling of radioisotopes, protection from chemical hazards, and containment of biohazardous materials. When working with any of these hazardous substances, proper eye protection should always be worn and mouth pipetting should never be employed.

RADIOISOTOPES

Radioisotopes commonly used for recombinant DNA applications are [^{3}H], [^{32}P], [^{35}S], and [^{125}I]. The half-lives of these isotopes, which are important for storage and disposal considerations, are 12.35 years, 14.3 days, 87.4 days, and 60 days, respectively.

Tritiated [^{3}H] DNA or RNA molecules are often used for *in situ* hybridization to tissue sections or chromosome preparations. Tritium is a soft ß-particle emitter, and this property minimizes scatter upon autoradiography of hybridized tissues so that silver grains form specifically over labeled structures. [^{35}S]-labeled DNA or RNA molecules can also be used for this purpose. In these molecules, one of the oxygen atoms of the α-phosphate group is replaced by a labeled sulfur atom. [^{35}S]-labeled molecules emit higher energy β particles than tritiated compounds and therefore autoradiographic exposure times are reduced. [^{35}S]-deoxynucleoside triphosphates are used with increasing frequency in Sanger dideoxynucleotide sequencing, where greater autoradiographic resolution is obtained than with higher energy [^{32}P]-labeled nucleosides. [^{32}P]-labeled compounds, however, afford reduced exposure times, and they are commonly used in Maxam–Gilbert sequencing and most filter hybridization experiments where highly labeled material is necessary for ready detection.

[^{125}I] is a γ-ray, x-ray, and electron emitter. [^{125}I]-labeled compounds used in conjunction with recombinant DNA technologies are usually proteins that have been conjugated with the radionuclide for rapid detection. Oftentimes, immunoglobulin molecules or *Staphylococcus aureus* protein A (which binds to some immunoglobulin molecules) are used as probes for detecting antibody–antigen complexes, such as those produced upon screening a recombinant DNA expression library.

The radiation safety organization at your institution is responsible for receiving radioactive materials, checking the vials for leakage, monitoring proper usage, and assuring proper disposal. Radioactive materials should be stored in a shielded container in a properly labeled refrigerator or freezer (as indicated by the radioisotope manufacturer). A log of material delivered and used must be maintained. Waste materials must be collected in leakproof containers, and solid and liquid wastes need to be separated. Waste for each type of radioisotope should be stored separately, since they are treated differently depending upon their half-lives, energies and nature of emission. Monitoring contamination of laboratory areas and personnel exposure levels can be accomplished by wipe tests with filters (followed by scintillation counting), use of body and ring badges, scans of personnel and laboratory surfaces with an appropriately sensitive Geiger–Muller survey meter (for [^{35}S] and [^{32}P]), a NaI crystal detector (for [^{125}I] compounds), and thyroid scans (for [^{125}I] compounds) or urinanalysis.

Individuals who use radioactive materials in the recombinant DNA laboratory must take appropriate precautions to minimize exposure levels, since carelessness can lead to unnecessary exposure levels. Pregnant women should be especially cautious regarding exposure and should consult their physician prior to use of radioisotopes. Personnel should wear laboratory coats, safety glasses, and plastic gloves. Smoking, eating, drinking, and application of cosmetics are prohibited in areas where radioactive samples are manipulated or stored, and these areas should be labeled with caution signs. Absorbent paper should be spread under the working area, and stock solutions should be opened over a tray to prevent spillage onto the laboratory bench or floor. No additional shielding precautions are required for [^{3}H]- and [^{35}S]-labeled materials, since their particle emissions are not of sufficient intensity to penetrate the dead layer of skin. However, a chemical fume hood should be used to prevent inhalation of any volatile labeled compounds. [^{32}P]- and [^{125}I]-labeled compounds are safely used behind Plexiglas shields (0.25 or 0.5 inches, respectively). Stock samples of [^{125}I] should be shielded by lead bricks in a fume hood. Use of thin lead shields with strong β-emitters (like [^{32}P]) allows the production of *bremsstrahlung*, i.e., collisions of β–particles with lead, resulting in X-ray emission. Their use should therefore be avoided. Plexiglas shields are available from commercial sources but most are manufactured in-house. Plans for a shield that prevents exposure to those working behind or on the side of the shielding are shown below. Note the presence of a Plexiglas insert within the shield. This permits the short-term storage of waste materials and prevents exposure of the researcher during subsequent manipulations. Blocks of Plexiglas with holes drilled partway through are useful for holding and transporting radioactive samples. Remember, minimize exposure times and maximize the distance and shielding between yourself and the labeled materials. See Zoon (1987) for additional information concerning safe handling of radioisotopes.

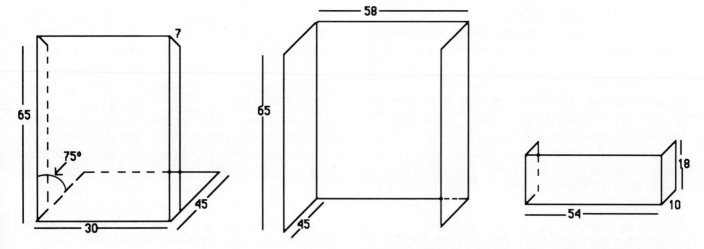

Plexiglas shield construction plans. Waste storage shield (on right) should be inserted into rear of side and back shielding (center). Front shielding (left) should then be inserted. Dimensions are in cm. Plexiglas should be at least 0.25 inches thick.

CHEMICAL HAZARDS

Chemical hazards include caustic substances that can injure tissues (such as acids, bases, and organic solvents) and mutagens (potential carcinogens) that can cause long-term genetic effects. Manufacturers of each chemical are required to provide (upon request) a Material Safety Data Sheet describing hazardous properties, handling and storage precautions, and decontamination procedures. In general, safety glasses, rubber gloves, and lab coats should be worn. Volatile compounds such as phenol, formaldehyde, chloroform, and ether should be used in a fume hood. Care should be taken to avoid breathing dust from materials such as ethidium bromide or chloramphenicol (possible carcinogens) or acrylamide (a neurotoxin). The reagents employed for DNA modification in the

Maxam–Gilbert DNA sequencing procedure are extremely hazardous and need to be properly used, stored, and inactivated. Waste materials should be collected in separate, labeled containers which are to be disposed of by the chemical safety office. When possible, toxic chemicals should be inactivated prior to storage as waste. Large volumes of flammable materials should be stored in safety cabinets, but never in non-explosion-proof refrigerators. Flammable chemicals should not be used near an open flame or another ignition source. Laboratories should be equipped with safety showers, eyewash stations, and fire extinguishers. Know their location.

BIOHAZARDS

The National Institutes of Health regulates the production and use of recombinant DNA molecules and organisms carrying these molecules. The guidelines are published in the Federal Register and are applicable to all research conducted in the United States at institutions which receive funding from the NIH. Other organizations are expected to voluntarily follow these rules. The NIH Office of Recombinant DNA Activities is responsible for setting the guidelines. Each organization has an Institutional Biosafety Committee that assures compliance. There are four biosafety levels of physical containment, and the level for a particular procedure is dependent upon the perceived safety risk as outlined in the guidelines. BL1 is the least stringent

containment level and BL4 is the most stringent. All levels require adherence to good microbiological practices, decontamination of work surfaces, use of mechanical pipetting devices, prohibition of eating, drinking, smoking, or makeup application, and decontamination of wastes prior to disposal. Higher levels of containment require more stringent conditions, such as biological safety cabinets, use of surgical masks, negative pressure rooms, etc. In addition to physical containment, various levels of biological containment are defined by the NIH guidelines. These depend on the viability of the host organism outside the laboratory environment as well as survival and transmission potentials of the vector from the host to another host.

All of the experiments described in this laboratory manual are exempt from the current NIH guidelines because they fall into one or more of the following categories: (1) they are not within organisms or viruses; (2) they consist entirely of DNA from a particular prokaryote and are propagated within that prokaryote; or (3) DNA is propagated using a λ-based or non-conjugative plasmid in an *E. coli* K-12 host that lacks conjugation-proficient plasmids or generalized transducing phages. Although permission to perform these experiments is not required from NIH or the Institutional Biosafety Committee, adherence to BL1 physical containment procedures, which are listed below, is recommended. The exemption described above should not be construed to indicate that manipulation of recombinant molecules in *E. coli* K-12 is not regulated by the NIH guidelines, since exceptions exist. These include molecular cloning of toxin genes and injection of recombinant DNA molecules into higher organisms.

Biosafety Level 1 Containment Procedures

- Follow standard microbiological practices.

- Access to the lab may be limited when experiments are in progress (at the discretion of the laboratory director).

- Decontaminate wastes prior to disposal. If they are to be moved prior to decontamination, they must be placed in a leakproof container.

- Mouth pipetting is prohibited.

- Eating, drinking, smoking, and cosmetic application is prohibited. Food may be stored in cabinets or refrigerators designated for this purpose only.

- A sink for hand-washing must be available within the laboratory.

- Wash hands prior to leaving the laboratory (after handling recombinant DNA-containing materials).

- Minimize the creation of aerosols.

- Labcoats are recommended.

- An insect and rodent control program must be in effect.

- Windows must be fitted with fly screens.

- The laboratory must be easily cleaned.

- Bench tops must be impervious to water and resistant to acids, alkalis, organic solvents, and moderate heat.

- Laboratory furniture must be sturdy.

REFERENCE

R. A. Zoon, *Methods Enzymol.* **152**, 25 (1987).

SOLUTIONS

CULTURE MEDIA

For culture plates, add 15 g of bacto agar per liter prior to autoclaving. For top agar, add 7.5 g of bacto agar. (Agarose should be substituted in procedures such as plaque lifts.)

❒ **LB broth** (per liter)
10 g of bacto-tryptone
5 g of bacto-yeast extract
10 g of NaCl
Adjust to pH 7.5 with NaOH or HCl

❒ **M9 salts** (per liter)
6 g of Na_2HPO_4
3 g of KH_2PO_4
0.5 g of NaCl
1 g of NH_4Cl

❒ **NZCYM** (per liter)
10 g of NZ amine (Type A hydrolysate of casein)
5 g of NaCl
5 g of yeast extract
1 g of casamino acids
2 g of $MgSO_4 \cdot 7H_2O$
Adjust to pH 7.5 with NaOH

❒ **2x YT** (per liter)
16 g of bacto tryptone
10 g of yeast extract
5 g of NaCl

BUFFERS AND OTHER REAGENTS

❒ **Ethidium bromide**
Stock of 10 mg/ml in water (store in light-protected container)
Dilute 1:10,000 in water (1 μg/ml) for gel staining

❒ **HTE buffer**
50 mM Tris–Cl, pH 8.0
20 mM EDTA

❒ **IPTG   (isopropyl-β-ᴅ-thiogalactoside)**
Dissolve in water to 100 mM
Sterilize by filtration through 0.2 μM Millipore filter

❑ **Loading buffer for agarose gel electrophoresis**
0.25% bromphenol blue
0.25% xylene cyanol
30% glycerol

❑ **Phenol**
Melt crystals of redistilled phenol by incubation at 50°C
Add 0.5 *M* Tris–Cl, pH 8.0, and shake
Add buffer until an aqueous layer is visible above the organic phase
Store refrigerated in a dark bottle
Discard if discoloration occurs

❑ **Pronase**
10 mg/ml in TNE buffer (preheat to 37°C for 15 minutes to digest any DNases present)

RESTRICTION ENZYME BUFFERS

❑ ***Bam*HI buffer (10x)**
1.5 *M* NaCl
60 m*M* Tris–HCl, pH 7.9
100 m*M* MgCl$_2$
1 mg/ml bovine serum albumin (BSA)

❑ ***Eco*RI buffer (10x)**
500 m*M* NaCl
900 m*M* Tris–HCl, pH 7.5
100 m*M* MgCl$_2$
1 mg/ml BSA

❑ ***Pst*I buffer (10x)**
500 m*M* NaCl
500 m*M* Tris–HCl, pH 8.0
100 m*M* MgCl$_2$
1 mg/ml BSA

❑ **RNase (pancreatic ribonuclease A)**
10 mg/ml in TE buffer (preheat to 80°C for 10 minutes to inactivate DNases)

❑ **SM buffer** (per liter)
5.8 g of NaCl
2 g of MgSO$_4$-7H$_2$O
50 ml of 1 *M* Tris–HCl, pH 7.5
5 ml of 2% gelatin

❑ **20x SSC** (per liter)
(20x = 3 *M* NaCl, 0.3 *M* Na citrate)
175.2 g of NaCl
88.2 g of Na citrate
Adjust to pH 7.0 with NaOH

❑ **TE buffer**
10 m*M* Tris–Cl, pH 8.0
0.1 m*M* EDTA

❑ **TEA (10x) for agarose gel electrophoresis** (per liter)
48.4 g of Trisma base
11.42 ml of glacial acetic acid
20 ml of 0.5 *M* EDTA, pH 8.0

❒ **TNE buffer**
10 mM of Tris–Cl, pH 8.0
10 mM of NaCl
0.1 mM EDTA

❒ **X-gal (5-dibromo-4-chloro-3-indolyl-β-D-galactoside)**
Dissolve in N, N-dimethylformamide to 2%
Wrap aluminum foil around the tube to avoid damage by light
Store in refrigerator

STRAINS AND DNA

E. coli strains

C600 [*F⁻ thi-1 thr-1 leuB6 lacY1 tonA21 supE44 λ⁻*]

HB101::Tn5 (Km^r) [F⁻, *hsdS20* (r$_B^-$, m$_B^-$) *recA13 ara-14 proA2 lacY1 galK2 rpsL20* (Sm^r) *xyl-5 mtl-1 supE44* λTn5 *kan* (Km^r)]

XL1-Blue [*recA1 endA1 gyrA96 thi hsdR17* (r$_k^-$, m$_k^+$) *supE44 relA1 lacZ* λ⁻], [F' *proAB lacI^{q}ZΔM15* Tn*10 tet* (Tc^r)] (Bullock *et al.*, 1987)

LE392 [*lacY galK2 galT22 metB1 trpR55 supE44 supF58 hsdR514* (r$_k^-$, m$_k^+$) λ⁻]

Note: *E. coli* strains, plasmids, and M13mp19 may be obtained from International Biotechnologies, Incorporated.

Phage stocks

M13mp19

Lambda phage recombinant library

Purified λ phage clone

Plasmids

pRSG192 ([*bla* (Ap^r), *V. harveyi chb*])

Note: pRSG192 described in Jannatipour et al. (1987) and Soto-Gil and Zyskind (1989).

pBR329 [*cat* (Cm^r), *tet* (Tc^r), *bla* (Ap^r)]

Note: pBR329 described in Covarrubias and Bolivar (1982).

DNA size markers

1-kb DNA ladder from BRL

Plasmid supercoiled ladder from BRL

REFERENCES

W. O. Bullock, J. M. Fernandez, and J. M. Short, *Biotechniques* **5**, 376 (1987).

L. Covarrubias and F. Bolivar, *Gene* **17**, 79 (1982).

M. R. Jannatipour, R. W. Soto-Gil, L. C. Childers, and J. W. Zyskind, *J. Bacteriol.* **169**, 3785 (1987).

R. W. Soto-Gil and J. W. Zyskind, submitted for publication (contains nucleotide sequence of *chb* gene; preprint available from J. W. Zyskind).

MOLECULAR BIOLOGY REAGENT SUPPLIERS

Amersham Incorporated
2636 South Clearbrook Drive
Arlington Heights, IL 60005
(312) 364-7100; (800) 323-9750

Bethesda Research Laboratories (BRL)
Life Techologies, Inc.
P.O. Box 6009
Gaithersburg, MD 20877
(301) 840-8000; (800) 638-8992

Bio-Rad Laboratories
2200 Wright Avenue
Richmond, CA 94804
(415) 232-7000; (800) 227-5589; (800) 645-3227

Boehringer Mannheim Biochemicals
9115 Hague Road
P.O. Box 50816
Indianapolis, IN 46250
(317) 576-2771; (800) 262-1640

Clontech Laboratories, Inc.
4055 Fabian Way
Palo Alto, CA 94303
(415) 424-8188; (800) 662-CLON

Du Pont NEN Research Products
549 Albany Street
Boston, MA 02118
(800) 551-2121

Fisher Scientific
711 Forbes Avenue
Pittsburg, PA 15219
(412) 562-8300

International Biotechnologies, Inc. (IBI)
275 Winchester Avenue
P.O. Box 9558
New Haven, CT 06535
(203) 562-3878; (800) 243-2555

New England BioLabs
32 Tozer Road
Beverly, MA 01915
(617) 927-5054; (800) 632-5227

Pharmacia/LKB Biotechnology, Inc.
800 Centennial Avenue
P.O. Box 1327
Piscataway, NJ 08854
(201) 457-8150; (800) 526-3593

Promega Corporation
2800 South Fish Hatchery Road
Madison, WI 53711
(608) 274-4330; (800) 356-9526

Stratagene
11099 North Torrey Pines Road
La Jolla, CA 92037
(619) 535-5400; (800) 424-5444

United States Biochemical Corporation (USB)
P.O. Box 22400
Clevelend, OH 44122
(216) 765-5000; (800) 321-0322

VWR Scientific
P. O. Box 7900
San Francisco, CA 94120
(415) 468-7150

FREQUENTLY USED ENZYMES

Enzyme	Activity	Use in Molecular Cloning
E. coli DNA polymerase I	1. Polymerizes dNTPs onto ds template with recessed 3' OH 2. 5'→3' exonuclease activity 3. 3'→5' exonuclease activity (ss DNA only—proofreading function)	Labeling DNA with radioactive dNTPs by "nick translation"
Klenow fragment of *E. coli* DNA polymerase I	Activities 1 and 3 listed above	1. Labeling 3' termini of DNA fragments by "fill-in" with radioactive dNTPs 2. Second-strand cDNA synthesis in cDNA cloning 3. Sequencing DNA using the Sanger dideoxy system
T4 DNA ligase	Links 5' phosphate to 3' OH	1. Closes nicks in DNA 2. Joins either cohesive or blunt ends generated by restriction enzymes
Bacterial alkaline phosphatase (BAP)	Catalyzes removal of 5' phosphate from DNA and RNA (heat-stable enzyme)	1. Removal of 5' phosphate from DNA or RNA for replacement by radioactive phosphate 2. Prevents "self-ligation" of DNA (e.g., prevents plasmids from recircularizing when treated with ligase since 5' phosphate is removed)
Calf intestinal phosphatase (CIP)	Catalyzes removal of 5' phosphate from DNA and RNA (inactivated by high temperatures)	Same as above, but more useful since it can be inactivated by heating prior to kinase labeling
T4 polynucleotide kinase	1. Transfers phosphate of 5' NTP to 5' OH of DNA or RNA 2. Transfers phosphate from 5' end of DNA or RNA to NDP 3. Removes 3' phosphoryl groups	1. 5' labeling of DNA or RNA by "forward reaction" after removal of 5' phosphate by phosphatase treatment 2. 5' labeling of DNA or RNA by "exchange reaction." Unlabeled phosphate transferred to NDP and replaced with labeled phosphate from labeled NTP
Terminal deoxynucleotidyl transferase	Adds deoxynucleotides to 3' OH end of DNA molecules	1. Adding complementary homopolymer tails to vector and cDNA 2. Labeling 3' ends of DNA fragments

Enzyme	Activity	Use in Molecular Cloning
T4 DNA polymerase	1. 5'→3' polymerase, polymerizes dNTPs onto ds template with recessed 3' OH 2. 3'→5' exonuclease activity	1. "Replacement-synthesis" for high level labeling of DNA. Exonuclease activity is allowed to remove nucleotides from 3' ends. Radiolabeled dNTPs are added to the reaction and are incorporated since polymerase activity is much much stronger than exonuclease. 2. 3' end labeling is accomplished by enzyme activity in presence of labeled dNTPs. 3' end is degraded until recessed 3'end is created. This is immediately filled in by polymerase.
Single stranded nucleases (S1, P1, and mung bean nucleases)	Cleaves ss DNA or RNA S1 cleaves across from nicks or gaps, mung bean nuclease cleaves only at gaps	1. Locate transcribed regions of DNA by hybridizing DNA to RNA and degrading ss DNA with ss nuclease 2. Locate introns in the same manner 3. Remove ss DNA regions of ds cDNA to allow cloning by homopolymer tailing or addition of linkers 4. Detect strand opening events during initiation of DNA replication
ExoVII (ss Exonuclease)	Degrades ss DNA and RNA at terminal regions. Does not cleave internal ss regions	In conjunction with S1 or mung bean nuclease, this enzyme is helpful in mapping the location of exons and introns
Bal 31	1. Degrades 3' and 5' strands of DNA in a progressive manner 2. Degrades ss DNA	1. Restriction mapping. DNA is progressively degraded and then cut with a restriction enzyme. Observation of restriction fragment disappearance indicates the order of the fragments 2. A set of deletions from a defined endpoint in a DNA clone can be generated by treatment with Bal 31
Bacteriophage SP6, T7, and T3 RNA polymerases	Recognizes SP6, T7, or T3 promoters and transcribes downstream sequences	In conjunction with radiolabeled NTPs, produces ss RNA transcripts for hybridization experiments Removal of the DNA template (with DNase) produces a probe which does not self- hybridize
Reverse transcriptase	Synthesizes copy DNA (cDNA) from ss DNA or RNA in the presence of a primer and dNTPs.	1. Makes DNA copies of ss DNA or RNA for cloning 2. Makes radiolabeled probes which are homologous to ss DNA or RNA

FREQUENTLY USED TECHNIQUES NOT DESCRIBED IN LABS

S1 NUCLEASE MAPPING

MATERIALS

- ❏ Yeast tRNA, 5 mg/ml
- ❏ 10x ss-hybridization buffer
  200 mM Tris–Cl, pH 8.0
  3 M NaCl
  1 mM EDTA
- ❏ 5x ds-hybridization buffer
  200 mM 4-(2-hydroxyethyl)-1-piperazineethanesulfonic acid (PIPES), pH 6.5
  2 M NaCl
  5 mM EDTA
- ❏ Deionized formamide
  Mix 25 ml of formamide with 2.5 g of mixed-bed, ion exchange resin (Bio-Rad AG 501-X8, 20–50 mesh). Stir for 30 minutes at room temperature. Filter twice through Whatman No. 1 filter paper. Dispense 1 ml aliquots into microcentrifuge tubes and store at –20°C.
- ❏ 10x S1 reaction buffer
  300 mM sodium acetate, pH 4.6
  500 mM NaCl

- 10 mM zinc sulfate
  50% glycerol
- ❏ 200 mM EDTA, pH 8.0
- ❏ Phenol:chloroform 1:1, equilibrated with 0.5 M Tris–Cl, pH 8.0
- ❏ 100% ethanol (–20°C)
- ❏ 80% ethanol, room temperature
- ❏ 100°C water bath
- ❏ Environmental shaker set at 63°C (single-stranded probe) or hybridization temperature (double-stranded probe)
- ❏ 7.5 M ammonium acetate
- ❏ Sterile, distilled water
- ❏ Ethanol dry ice bath
- ❏ Vacuum centrifuge
- ❏ Microcentrifuge
- ❏ Loading buffer
  8 M urea
  0.1% xylene cyanol
  0.1% bromophenol blue
- ❏ 8% acrylamide–8% urea sequencing gel (see Lab VII, Day 4)

PROTOCOL

1. Mix the single-stranded (ss) or double-stranded (ds) probe (at least 50,000 cpm) with 150 µg total RNA. When using less than 150 µg of RNA, the final concentration of RNA should be brought to 150 µg with yeast tRNA.
2. Lyophilize the mixture in a vacuum centrifuge.
3. For ss DNA probes:
   a. Resuspend in 27 µl sterile water: mix with a pipetter, vortex, centrifuge briefly to move liquid to bottom of tube.
   b. Add 3 µl 10x ss-hybridization buffer. Heat the tubes at 100°C for 2 minutes.
   c. Without cooling, transfer to an environmental shaker set at 63°C. Hybridize for 3 hours.
   d. Go to step 5.
4. For ds DNA probes:
   a. Resuspend pellet in 6 µl of 5x ds-hybridization buffer.
   b. Add 24 µl deionized formamide (final concentration is 80%), mix with a pipetter, vortex, then centrifuge briefly to move liquid to bottom of tube.
   c. Heat in a water bath at 75°C for 15 minutes to denature the dsDNA probe.
   d. Without cooling, transfer to an environmental shaker set at the temperature of hybridization for 5 to 6 hours without shaking. The hybridization temperature is calculated by adding 5 to 8°C to the result obtained from the following formula:

$$81.5 + 16.6 \times \log[Na^+] + 0.41[\%GC] - 0.6[\% \text{ formamide}] - 500 \text{ bp}/\# \text{ of bp in probe}$$

   e. Go to step 5.
5. Add 150 units S1 nuclease, 30 µl of 10X S1 reaction buffer, and water to 300 µl. Mix gently and rinse the sides of the tube with the pipetter. Incubate at 37°C for 30 minutes.
6. Terminate the reaction with 4 µl of 200 mM EDTA.
7. Add an equal volume of phenol:chloroform, vortex briefly (a few seconds) to mix, and centrifuge at full speed in a microcentrifuge for three minutes to separate the phases. Carefully remove the *top* layer with a sterile pasteur pipette, and add to a new microcentrifuge tube.
8. Add 10 µg of yeast tRNA, 120 µl of 7.5 M ammonium acetate, and 700 µl 100% ethanol. Mix well by inverting the tube several times, and freeze in an ethanol-dry ice bath for 15–30 minutes. Centrifuge for 15 minutes in a microcentrifuge. Decant the ethanol.

Note: To remove the ethanol without disturbing the pellet, use a pasteur pipette with a thinned end: melt the glass of a pasteur pipette, pull out to thin the width of the end, break off. Use such a pipette to remove the ethanol in all further steps.

9. Wash the RNA pellet with 1.0 ml of 80% ethanol at room temperature, vortex briefly, spin for 5 minutes, and pour off supernatant. Repeat 2 times.
10. Dry under vacuum, resuspend in 40 µl sterile distilled water, and lyophilize.
11. Dissolve pellet in 5 µl of loading buffer, and load 1.2 to 2.0 µl of the sample onto a sequencing gel. Include labeled molecular weight markers or a sequencing ladder of the probe containing the A and A + G sequencing reactions of Maxam and Gilbert (1980) so that the size of the protected probe fragment can be estimated.

REFERENCES

H. Aiba, S. Adhya, and B. de Crombrugghe, *J. Biol. Chem.*, **256**, 11905-11910 (1981).

G. A. Kassavetis, T. Elliott, D. P. Rabussay, and E. P. Geiduschek, *Cell*, **33**, 887-897 (1983).

A. M. Maxam, and W. Gilbert, *Meth. Enzym.*, **65**, 499-560 (1980).

J. Casey, and N. Davidson, *Nuc. Acids Res.*, **4**, 1539-1552 (1977).

PRIMER EXTENSION

MATERIALS

❒ End-labeled probe for extension (double stranded or single stranded)
❒ RNA sample
❒ Microcentrifuge tubes (sterile)
❒ Microcentrifuge
❒ 5x hybridization salts

200 mM PIPES pH 6.5
2 M NaCl
5mM EDTA
❒ Deionized formamide
Mix 5 ml of formamide and 0.5 g of mixed-bed, ion-exchange resin (Bio-Rad 501-X8). Stir for 30 minutes at room temperature. Filter twice through Whatman No. 1 filter paper. Dispense into 1 ml aliquots and store at –20°C.
❒ 7.5 M ammonium acetate
❒ 100% ethanol
❒ Dry ice-ethanol bath
❒ 80% ethanol
❒ Vacuum drying apparatus
❒ 0.5 M EDTA, pH 8.0
❒ Avian myeloblastosis virus reverse transcriptase
❒ Extension Mix
0.05 M Tris–HCl, pH 8.0
0.01 M MgCl$_2$
0.06 M NaCl
5mM dithiothreitol
1.5 mM each of dATP, dCTP, dGTP and TTP.
(One or more of the deoxynucleoside triphosphates may be radiolabeled if necessary).
❒ Loading buffer
8 M Urea
0.1% xylene cyanol
0.1% Bromophenol blue

PROTOCOL

1. Lyophilize RNA sample (150 µg) with the probe (at least 35,000 cpm) in a sterile microcentrifuge tube.
2. Resuspend in 6 µl of 5x hybridization salts. Vortex well and centrifuge briefly.
3. Add 24 µl of deionized formamide. Vortex well and centrifuge briefly.
4. Heat at 75°C for 15 minutes to denature the probe.
5. Let cool down progressively to the temperature of hybridization which can be calculated based upon the G–C content of the hybrid (see S1 nuclease mapping protocol).
6. Hybridize for 5-6 hours at the temperature of hybridization.
7. Ethanol precipitate by bringing the volume up to 100 µl with sterile distilled water. Add 80 µl of 7.5 M ammonium acetate and 460 µl of 100% ethanol. Invert several times. Precipitate sample in a dry ice-ethanol bath for 15 minutes.
8. Centrifuge for 15 minutes in the microcentrifuge. Carefully discard the supernatant without disturbing the pellet.
9. Wash the pellet with 80% ethanol (room temperature). Vortex briefly.
10. Centrifuge for 5 minutes. Carefully discard the supernatant without disturbing the pellet. Dry the pellet *in vacuo*.
11. Resuspend pellet in 20 µl of Extension Mix. Add 10 units of reverse transcriptase.
12. Incubate at 37°C for 90 minutes.
13. Stop the reaction by addition of 2 µl of 0.5 M EDTA.
14. Add 8 µl of loading buffer and load half of the total volume (30 µl) on a denaturing polyacrylamide gel along with size markers. Locate the extended product by autoradiography.

REFERENCES

H. Aiba, S. Adhya, and B. de Crombrugghe, *J. Biol. Chem.*, **256**, 11905-11910 (1981).

J. Casey, and N. Davidson, *Nucleic Acids Res.* **4**, 1539-1552 (1977).

H. Serizawa and R. Fukuda, *Nucleic Acids Res.* **15**, 1153-1163 (1987).

RNA ISOLATION FROM BACTERIAL CELLS

MATERIALS

❐ An overnight culture of a bacterial strain grown in M9 medium or a minimal medium that supports growth of the particular bacterial species
❐ 500 ml M9 medium
❐ Sorvall GS3 rotor and bottles (500 ml)
❐ Sorvall SS34 rotor
❐ Polypropylene centrifuge tubes
❐ Sorvall Superspeed centrifuge
❐ TNE
❐ RNA lysis buffer:
  20 mM sodium acetate, pH 5.5
  0.5% SDS
  1 mM EDTA

❒ Phenol saturated with 20 m*M* sodium acetate
❒ 95% ethanol (–20°C)
❒ 95% ethanol, room temperature
❒ 60°C water bath
❒ 0.3 *M* sodium acetate
❒ Sterile, distilled water
❒ Ethanol–dry ice bath
❒ Microcentrifuge
❒ Vacuum centrifuge

PROTOCOL

1. Add 5 ml of the overnight culture to 500 ml of M9 broth.
2. Grow culture to OD_{450} of 0.3 to 0.5.
3. Divide the culture into 2 centrifuge bottles, and centrifuge cells for 10 minutes at 6000 rpm using the GS3 rotor.
4. Resuspend the cells in 10 ml TNE, and transfer to a 15-ml polypropylene tube.
5. Centrifuge in an SS34 rotor at 5000 rpm for 5 minutes.
6. Resuspend the cells in 3 ml RNA lysis buffer. Add 3 ml of hot phenol (60°C), and shake gently at 60°C.
7. Centrifuge in an SS34 rotor for 5 minutes at 3000 rpm.
8. Transfer the aqueous phase (top) to a new polypropylene tube, and repeat the phenol extraction once more at 60°C and an additional two times at room temperature.

9. Precipitate the RNA with 9 ml of 95% ethanol (–20°C). Mix well, and freeze in an ethanol-dry ice bath for 30 minutes. Centrifuge for 45 minutes in an SS34 rotor. Decant the ethanol.
10. Dissolve the pellet in 250 μl 0.3 *M* sodium acetate and repeat precipitation with 95% ethanol. Repeat one more time so that the sample has been precipitated three times with ethanol.
11. Wash RNA pellet with 1.0 ml of 95% ethanol, vortex very briefly, spin for 5 minutes, and pour off supernatant. Repeat 2 times.
12. Dry pellet in a vacuum centrifuge, and resuspend in 1 ml sterile distilled water.
13. Measure OD_{260} (1 OD_{260} is equivalent to 40 μg/ml of RNA).
14. Store RNA in 2 volumes of 95% ethanol at –70°C.

REFERENCES

H. Aiba, S. Adhya, and B. de Crombrugghe, *J. Biol. Chem.* **256**, 11905-11910 (1981).

RNA ISOLATION FROM ANIMAL CELLS

MATERIALS

❒ Solution 1:
　6 *M* urea
　3 *M* LiCl (do not autoclave)
❒ Solution 2:
　10 m*M* Tris–Cl, pH 7.5

10 mM EDTA

1% SDS (autoclave)

❐ Tissue homogenizer (sterile)

❐ Microcentrifuge tubes (sterile)

❐ Bucket of ice

❐ Microcentrifuge at 4°C

❐ Phenol equilibrated with TE.

❐ Chloroform

❐ Sterile micropipette tips

❐ Sterile 5 M NaCl

❐ Ethanol–dry ice bath

❐ Sterile TE

PROTOCOL

1. Homogenize about 50 mg of tissue in 1 ml of solution 1.
2. Transfer sample into a microcentrifuge tube
3. Rinse homogenizer with 500 µl solution 1 and add rinse to microcentrifuge tube.
4. Put sample on ice for at least 4 hours to precipitate the RNA (DNA is soluble in 3 M LiCl).
5. Centrifuge for 10 minutes at 4°C in a microcentrifuge. Discard supernatant.
6. Resuspend pellet in 300 µl of solution 2.
7. Add 300 µl phenol, vortex for 5 seconds.
8. Add 300 µl chloroform, vortex for 5 seconds.
9. Centrifuge for 5 minutes in microcentrifuge.
10. Remove upper (aqueous) phase to new microcentrifuge tube.
11. Repeat steps 7–10.

12. Add 300 µl chloroform. Vortex for 5 seconds and centrifuge for 5 minutes.
13. Add 20 µl of 5 M NaCl to aqueous phase. Mix. Add 1 ml ethanol and freeze at –70°C (ethanol–dry ice bath) for 20 minutes.
14. Centrifuge for 10 minutes at 4°C in a microcentrifuge. Discard supernatant.
15. Dry samples under vacuum.
16. Dissolve the pellet in 50 µl of sterile TE. Keep samples on ice and freeze at –70°C for storage. OD_{260} measurements may be made to determine amount of material recovered (1 OD_{260} = 40 µg/ml of RNA).

ELECTROPHORESIS OF RNA IN A DENATURING FORMALDEHYDE-AGAROSE GEL; RNA GEL BLOTTING (NORTHERN BLOTTING)

MATERIALS

❐ Agarose

❐ 1x sodium phosphate buffer pH 7.0 (10 mM)

❐ Horizontal gel frame

❐ Comb

❐ RNase-free electrophoresis apparatus (submerged type)

❐ Peristaltic pump and tubing

❐ Power supply

❐ Formaldehyde (37%)

❐ RNA loading buffer

16 µl 1 M sodium phospate buffer, pH 7.0

100 μl H$_2$O
60 μl bromphenol blue (10 mg/ml)
260 μl formaldehyde (37%)
720 μl deionized formamide
100 μl 80% glycerol
10 μl ethidium bromide (10 mg/ml)
- ☐ UV transilluminator and camera set-up
- ☐ 5 m*M* NaOH
- ☐ nylon membrane (such as Zeta-Probe from Bio-Rad or Genescreen plus from NEN)
- ☐ 2x SSC

PROTOCOL

1. Melt agarose by boiling in 1x phosphate buffer. A good agarose concentration is 1.5%.
2. Cool down to 65°C and add 6.5 ml of formaldehyde (37%) for every 100 ml of gel solution.
3. Pour gel in fume hood, cover with plastic box to prevent evaporation. Let harden for 30 minutes to an hour.
4. Fill electrophoresis apparatus with 10 m*M* sodium phosphate buffer.
5. Prepare samples by adding up to 7 μl (10 μg) of RNA in sterile H$_2$O to 40 μl of RNA loading buffer.
6. Heat for 3 minutes at 65°C to denature RNA.
7. Load samples and apply 100 volts for 3–4 hours (or 25 volts overnight). Recirculate buffer with pump to prevent pH gradient from forming.

8. Take a photograph of the gel on a UV transilluminator prior to blotting.
9. Use a razor blade to remove extraneous portions of the gel.
10. Soak the gel for 30 minutes in 5 m*M* NaOH. Repeat once.
11. Transfer as per Southern blot procedure to a nylon membrane using 5m*M* NaOH as the blotting solution. Allow transfer to proceed for at least 6 hours.
12. Remove the filter from the gel and rinse for 30 seconds with 2x SSC.

Caution: Longer times may result in RNA removal.

13. Bake filter in a vacuum oven for 2 hours at 80°C. Hybridization is performed as for DNA gel blots.

LABELING DNA AT THE 3' END BY THE FILL-IN REACTION WITH KLENOW FRAGMENT

MATERIALS
- ☐ Restriction fragment containing a 5' overhang
- ☐ [α-^{32}P]dATP, 10 mCi/ml, 3000 Ci/mmole
- ☐ [α-^{32}P]dCTP, 10 mCi/ml, 3000 Ci/mmole
- ☐ [α-^{32}P]dGTP, 10 mCi/ml, 3000 Ci/mmole
- ☐ [α-^{32}P]TTP, 10 mCi/ml, 3000 Ci/mmole
- ☐ Klenow fragment of DNA polymerase I
- ☐ Klenow fragment 10x buffer, (use manufacturer's recommendation)

❑ Mixture of unlabeled dATP, dCTP, dGTP, and TTP (20 m*M*)

❑ Phenol:chloroform 1:1, equilibrated with 0.5 *M* Tris–Cl, pH 8.0

❑ Ether

❑ 95% ethanol (-20°C)

❑ 70% ethanol, room temperature

❑ 37°C water bath

❑ 3.0 *M* sodium acetate

❑ Sterile, distilled water

❑ Column containing Biorad P-30 resin equilibrated in TE

❑ Ethanol–dry ice bath

❑ Microcentrifuge

Note: The 1 kb and 123 bp ladders from BRL can be labeled with the same protocol.

PROTOCOL

1. Digest 5–15 µg of DNA with a restriction enzyme that will generate 3'recessed ends (5' overhanging ends).

2. Run a small agarose gel to insure that digestion is complete.

3. Add an equal volume of phenol:chloroform to the DNA sample, vortex briefly (a few seconds) to mix, and centrifuge at full speed in a microcentrifuge for three minutes to separate the phases. Carefully remove the *top* layer with a sterile pasteur pipette, and add to a new microcentrifuge tube.

4. Extract with an equal volume of ether, vortex briefly, and discard *top* layer in waste receptacle.

5. Add 3.0 *M* sodium acetate to make the solution 0.3 *M*.

6. Precipitate the DNA with 2.5 volumes of 95% ethanol (-20°C). Mix well and freeze in an ethanol dry ice bath for 15-30 minutes. Centrifuge for 15 minutes in a microcentrifuge. Decant the ethanol.

7. Wash DNA pellet with 1.0 ml of 70% ethanol at room temperature, vortex briefly , centrifuge for 5 minutes, and pour off supernatant. Repeat 2 times.

8. Dry under vacuum and resuspend in 20 µl sterile distilled water.

9. Labeling reaction: Use only those nucleotides that are complementary to the nucleotides in the 5' overhang; replace volumes with water if any nucleotides are left out:
   20 µl DNA
   10 µl [α-^{32}P]dATP
   10 µl [α-^{32}P]dCTP
   10 µl [α-^{32}P]dGTP
   10 µl [α-^{32}P]TTP
   10 µl Klenow fragment 10x buffer
   30 µl water (+ 10 µl for missing nucleotides)
   10 units of Klenow fragment

10. Incubate at room temperature for 30 minutes.

11. Chase for 10 minutes with 5 µl of a mixture of deoxynucleoside triphosphates (2 m*M* each).

12. Remove the unincorporated triphosphates by passing the DNA through a column containing

Biorad P-30 resin equilibrated in 10 mM Tris–Cl, 1 m*M* EDTA (pH 8.0) as described in Lab XIII, Day 2.

13. Phenol extract, ethanol precipitate, and dry the pellet as described in steps 3 through 8.
14. Resuspend the probe in water.
15. Digest the labeled DNA with a second restriction enzyme to obtain fragments labeled only at one end.
16. Isolate the restriction fragment on an agarose or acrylamide gel and determine the radioactivity by scintillation counting.

LABELING DNA AT THE 5' END WITH POLYNUCLEOTIDE KINASE

MATERIALS

❑ Restriction fragment containing 5' overhangs, 3' overhangs, or blunt ends
❑ [γ-^{32}P]ATP, 5000 Ci/mmole, 10 mCi/ml
❑ Calf intestinal phosphatase (CIP)
❑ Diethyl pyrocarbonate (DEPC) freshly diluted 1:10 in 95% ethanol, (keep on ice)
❑ 10x CIP stop buffer
   100 mM Tris–Cl pH 8.0
   3.0 *M* sodium acetate
   10 mM EDTA
❑ 95% ethanol (–20°C)
❑ 70% ethanol, room temperature
❑ 37°C water bath

❑ 10x kinase buffer type I:
   0.5 *M* Tris–Cl, pH 7.6
   0.1 *M* MgCl$_2$
   50 mM dithiothreitol
   1 mM spermidine
   1 mM EDTA
❑ 10x kinase buffer type II:
   0.2 *M* Tris–Cl, pH 9.5
   10 mM spermidine
   1 mM EDTA
❑ 10x kinase buffer type III:
   0.5 *M* Tris–Cl, pH 9.5
   0.1 *M* MgCl$_2$
   50 mM dithiothreitol
   50% glycerol
❑ T4 polynucleotide kinase
❑ 5.0 *M* ammonium acetate
❑ Sterile, distilled water
❑ Ethanol–dry ice bath
❑ Microcentrifuge

PROTOCOL

1. Digest 5–30 μg of DNA with a restriction enzyme in the presence of calf intestinal phosphatase (CIP) as described in Lab VI, Days 3 and 4.
2. Run a small agarose gel to insure that digestion is complete.
3. For 5' protruding ends,
   a. Dissolve pellet in 90 μl water, 10 μl 10x kinase buffer I, 100 μCi [γ-^{32}P]ATP, and 20 units T4 polynucleotide kinase.

b. Mix and incubate at 37°C for 1 hour, then proceed to step 5.

4. For 5' recessed and blunt ends:
   a. Dissolve pellet in 36 µl water and 4 µl 10x kinase buffer II.
   b. Heat at 70°C for 2 minutes. Chill quickly in ice water bath.
   c. Add 5 µl 10x kinase buffer III, 100 µCi [γ-^{32}P]ATP, and 20 units T4 polynucleotide kinase.
   d. Mix and incubate at 37°C for 1 hour. Proceed to step 5.

5. Precipitate the DNA by addition of 5 µl $5M$ ammonium acetate and 400 µl 95% ethanol.

6. Mix well and freeze in an ethanol dry ice bath for 15–30 minutes. Centrifuge for 15 minutes in a microcentrifuge. Decant the ethanol.

7. Wash DNA pellet with 1.0 ml of 70% ethanol at room temperature, vortex briefly, centrifuge for 5 minutes, and pour off supernatant. Repeat 2 times.

8. Dry under vacuum, and resuspend in 20 µl sterile distilled water.

9. Digest the labeled DNA with a second restriction enzyme to obtain fragments labeled only at one end.

10. Isolate the restriction fragment on an agarose or acrylamide gel and determine the radioactivity by scintillation counting.

REFERENCES

A. M. Maxam, and W. Gilbert, *Meth. Enzym.* **65,** 499-560 (1980).

T. Maniatis, E. F. Fritsch, and J. Sambrook, "Molecular Cloning—A Laboratory Manual," Cold Spring Harbor Lab., Cold Spring Harbor, New York (1982).

TRANSDUCTION WITH P1 BACTERIOPHAGE

MATERIALS

❑ MC buffer:
   100 mM MgSO$_4$
   50 mM CaCl$_2$
❑ 1x M9 salts (salts in M9 medium) containing 0.1M sodium citrate
❑ R agar plates freshly made the day of use (do not dry plates)
   10 g Bacto tryptone
   1 g yeast extract
   8 g NaCl
   12 g agar
   1 liter water
   Autoclave, and after cooling add 2 ml of 1 M CaCl$_2$ and 5 ml of 20% glucose
❑ R-Top Agar: (same as R agar except add 5g/l of agar)
❑ L broth containing 5 mM CaCl$_2$
❑ 1 M sodium citrate
❑ Chloroform
❑ P1vir stock

❑ P1-sensitive *E. coli* strain, *e.g* ., LE392
❑ Oakridge tubes
❑ Sorvall Superspeed centrifuge
❑ Sorvall SS34 rotor
❑ Microcentrifuge tubes
❑ Microcentrifuge

PROTOCOL

A. Preparation of Lysates:

1. Prepare a fresh overnight culture of your donor strain in L broth containing 5 mM CaCl$_2$.
2. Make a 1:100 dilution of the overnight culture into L broth containing 5 mM CaCl$_2$. The calcium is necessary for adsorption of phage to cells.
3. Aerate at 37°C until an OD$_{450}$ of 0.4 (~2 x 10^8 cells/ml) is reached.
4. Prepare dilutions of a P1*vir* stock and mix 100 µl of P1 dilutions with 900 µl of cells; (about10^7 phage per 1 ml of culture is optimum):
   a. 100 µl of P1 stock (if titer is low) + 900 µl of cells
   b. 100 µl of (1:10) P1 stock dilution + 900 µl of cells
   c. 100 µl of (1:100) P1 stock dilution + 900 µl of cells
   d. 100 µl of (1:1000) P1 stock dilution + 900 µl of cells
5. Incubate with mixing at 37°C for 20 minutes.

6. Mix each sample with 2.5 ml of melted R-top agar held at 45°C. Mix tube by rolling back and forth between palms of your hands so as to minimize bubbles. Pour onto a freshly made R agar plate. Higher titers are obtained with plates poured the day of use.
7. Incubate plates right side up for about 6 to 8 hours; watch plates to make sure that lysis has occurred before proceeding to next step.
8. Add 1 ml L broth containing 5 mM CaCl$_2$ to each plate that has observable lysis.
9. Scrape off the top agar layer into an Oakridge tube.
10. Add 300 µl of chloroform, mix by vortexing, and let incubate at room temperature for 10 minutes.
11. Centrifuge in an SS34 rotor at 10,000 rpm for 10 minutes.
12. Transfer the supernatant containing the P1 lysate to a sterile test tube, and add 300 µl of chloroform.
13. Mix, and store at 4°C

B. Titration of lysates by determining the number of plaque forming units per ml (pfu/ml):

1. Dilute an overnight culture of P1-sensitive *E. coli* 1:20 into L broth containing 5 mM CaCl$_2$. Aerate for 2 hours at 37°C.
2. Prepare serial dilutions of P1 in L broth containing 5 mM CaCl$_2$. For example,

a. 5 µl of P1 stock + 5 ml broth (10^{-3})

b. 5 µl of 10^{-3} dilution + 5 ml broth (10^{-6})

c. 100 µl of 10^{-6} dilution + 900 µl broth (10^{-7})

d. 100 µl of 10^{-7} dilution + 900 µl broth (10^{-8})

3. Mix 100 µl of each of the last three dilutions with 100 µl of cells. Shake at 37°C for 15 minutes.

4. Mix with 2.5 ml of melted R-top agar. Pour on R agar plate.

5. Look at plates after 8 to 12 hours, count plaques, and calculate pfu in original P1 stock.

C. Transduction with P1 lysates:

1. Prepare an overnight culture of recipient cells.

2. Dilute 5 ml of the overnight culture with 30 ml of the L broth containing 5 mM $CaCl_2$, and shake at 37°C for about 30–45 minutes.

3. Transfer cells to an Oakridge tube and centrifuge in an SS34 rotor at 10,000 rpm for 10 minutes.

4. Resuspend pellet in 5 ml MC buffer.

5. Mix in microcentrifuge tubes:

a. 100 µl of P1 stock + 0.9 ml of cells

b. 100 µl of (1:10) P1 stock dilution + 0.9 ml of cells

c. 100 µl of (1:100) P1 stock dilution + 0.9 ml of cells

d. 0.9 ml of cells alone

6. Shake gently at 37°C for 20 minutes.

7. Add 500 µl of 1 M sodium citrate to each microcentrifuge tube.

8. Centrifuge 30 seconds in microcentrifuge. Pour off supernatant.

9. Resuspend pellet in 500 µl of M9 salts containing 0.1 M sodium citrate.

10. Spread 100 µl of cells on selective plates. Centrifuge briefly again, resuspend pellet in 100 µl, and spread the rest of the cells on selective plates. Incubate at 37°C.

REFERENCE

J. H. Miller, "Experiments in Molecular Genetics," Cold Spring Harbor Lab., Cold Spring Harbor, New York (1982).

LARGE-SCALE PLASMID ISOLATION BY THE CLEARED LYSATE METHOD

Day 1 : Isolation of a clone of cells

MATERIALS

❏ Cells of choice (from a glycerol stock or agar vial)

❏ LB agar plates with appropriate antibiotic (depending on the resistance marker of the plasmid)

❏ Inoculating loop

❏ 37°C incubator

PROTOCOL

1. Streak cells on agar LB plates

2. Grow overnight at 37°C

Day 2 : Preparation of overnight culture

MATERIALS
- ❏ Plates with streaked cells
- ❏ LB broth containing antibiotic for selection of plasmid-containing cells
- ❏ Shaker-incubator at 37°C
- ❏ 15-ml sterile-capped tube

PROTOCOL
1. Inoculate 5 ml LB broth (plus antibiotic) with a single colony
2. Shake overnight at 37°C

Day 3 : Preparation of large culture and plasmid amplification

MATERIALS
- ❏ LB broth containing antibiotic for selection of plasmid-containing cells
- ❏ Shaker-incubator at 37°C
- ❏ Sterile 2-L flask
- ❏ Spectrophotometer
- ❏ Chloramphenicol (150 mg/ml in ethanol) or spectinomycin (50 mg/ml in water)

PROTOCOL
1. Add 1 ml of overnight culture into 500 ml of LB broth (plus antibiotic) in a 2-L flask.
2. Let cells grow for about 3 hours shaking at 37°C until they reach an OD_{450} of 0.8–1.0.

3. Add 1 ml of chloramphenicol (150 mg/ml in ethanol) to a final concentration of 150 µg/ml, and shake overnight at 37°C to amplify the plasmid. Use spectinomycin at 50 µg/ml if cells are resistant to chloramphenicol.

Day 4 : Isolation of Plasmid DNA

MATERIALS
- ❏ Sorvall GS3 rotor and bottles (500 ml)
- ❏ Sorvall SS34 rotor
- ❏ Sorvall Superspeed centrifuge
- ❏ TNE
- ❏ Oakridge tubes
- ❏ Sucrose buffer
  25% sucrose,
  0.25M Tris–Cl, pH 8.0
- ❏ Lysozyme solution (10 mg/ml in TNE)
- ❏ 0.5 M EDTA, pH 8.0
- ❏ Lysis buffer
  1% Brij-58,
  0.4% deoxychlolate,
  1% Triton X-100,
  63 mM EDTA,
  50 mM Tris–Cl, pH 8.0)
- ❏ Triton X-100
- ❏ Container of ice
- ❏ Sterile graduated cylinder (25–50 ml)
- ❏ Molecular biology grade cesium chloride
- ❏ Ethidium bromide solution (10 mg/ml)

❐ Sorvall TV850 ultracentrifuge tubes (or equivalent)
❐ CsCl in TNE
  prepare by adding 1 gram of CsCl per ml of solution
❐ Sorvall TV850 ultracentrifuge rotor (or equivalent)

Note: Make fresh lysozyme solution each time.

PROTOCOL

1. Divide the culture into two GS3 bottles and centrifuge at 9000 rpm at 4°C for 5 minutes in the GS3 rotor. At this point you can discard the supernatant and freeze the pellet at –20°C if you wish to continue the experiment another day.
2. Resuspend each pellet in 10 ml of TNE. Transfer combined sample to an Oakridge tube.
3. Centrifuge at 7000 rpm for 7 minutes in an SS34 rotor.
4. Remove the supernatant. Resuspend the pellet thoroughly in 3.8 ml of sucrose buffer. Mix well and place on ice for 5 minutes.
5. Add 0.75 ml of lysozyme solution. Mix gently and put on ice for 5 minutes.
6. Add 0.75 ml of 0.5 M EDTA. Place on ice for 5 minutes.
7. Add 6 ml of lysis buffer. Mix gently.
8. Place on ice for about 5 minutes until the solution becomes viscous. If lysis does not work well, add 250 µl of Triton X-100.

9. Centrifuge at 15,000 rpm at 4°C for 45 minutes in an SS34 rotor.
10. Carefully decant supernatant into a sterile graduated cylinder. Lysates from 1 liter of cells will fit into a TV850 ultracentrifuge tube. Add TNE to a volume of 25 ml.
11. Add 1 g of CsCl per 1 ml of supernatant. The sample may be stored at this stage prior to the addition of ethidium bromide and centrifugation.
12. Add 0.8 ml of ethidium bromide solution (10 mg/ml) for every 10 ml of sample.
13. Mix thoroughly, and transfer to TV850 tubes. Remove any bubbles, and add CsCl/TNE solution to fill the tube up to the neck. Balance tubes so that they have equivalent weights by the addition of CsCl/TNE solution.
14. Load ultracentrifuge rotor, and centrifuge at 45000 rpm for 16–24 hours at 20°C.

Day 5: Isolation of the plasmid DNA band and further purification by high speed centrifugation.

MATERIALS
❐ Ringstand with clamp
❐ 18-gauge needles
❐ 10-ml syringes
❐ Ultraviolet light
❐ Sorvall TV865 ultracentrifuge tubes (or equivalent)
❐ CsCl in TNE (prepare by adding 1 gram of CsCl per ml of solution)

❏ Sorvall TV865 ultracentrifuge rotor (or equivalent)

❏ 3 *M* sodium acetate
❏ Cold 100% ethanol

PROTOCOL

1. Remove centrifuge tube from rotor and clamp to a ringstand. Locate the plasmid band using a UV light. The supercoiled plasmid is unable to bind as much ethidium bromide as the nicked or linear DNA. Because ethidium bromide is less dense than DNA, nicked and linear DNA should be above the plasmid DNA.
2. Remove the band with an 18-gauge needle attached to a 5-ml syringe. (Be sure to punch a hole in the top of the tube first if it is sealed).
3. Transfer the sample to a TV865 centrifuge tube, balance tubes with CsCl/TNE, and centrifuge again at 45,000 rpm for 16–24 hours at 4°C.

Day 6: Removal of plasmid band and extraction of DNA from CsCl/ethidium bromide solution

MATERIALS
❏ Ringstand with clamp
❏ 18-gauge needles
❏ 10-ml syringes
❏ Sterile glass test tubes to hold 20 ml
❏ Isoamyl alcohol
❏ Ultraviolet light
❏ TNE
❏ Oakridge tubes

PROTOCOL

1. Locate plasmid band with UV light and remove with an 18-gauge needle attached to a syringe.
2. Transfer to a sterile glass tube, and extract the ethidium bromide by adding an equal amount of isoamyl alcohol. Vortex.
3. Remove the upper alcohol layer with a Pasteur pipette and discard. Repeat extractions until the pink color is gone.
4. Add 5 volumes of TNE and 3*M* sodium acetate to make the solution 0.3 *M*. Mix well, and transfer to an Oakridge tube.
5. Precipitate the DNA by adding 2.5 volumes of cold 100% ethanol and mix gently by inverting. Store overnight at –20°C (or in an ethanol–dry ice bath for 15 minutes).

Day 7: Pelleting the purified DNA

MATERIALS
❏ Sorvall SS34 rotor
❏ Sorvall Superspeed centrifuge
❏ 0.3 *M* sodium acetate
❏ Microcentrifuge tubes
❏ Cold 100% ethanol
❏ 70% ethanol
❏ Ethanol–dry ice bath
❏ Microcentrifuge

❐ Vacuum drying apparatus
❐ TNE

PROTOCOL

1. Centrifuge precipitated DNA for 40 minutes at 15,000 rpm in an SS34 rotor at 4°C.
2. Carefully remove the supernatant without disturbing the pellet. Drain the residual ethanol on a paper towel.
3. Dissolve the pellet in 250 µl of sodium acetate and transfer DNA to a microcentrifuge tube. You may wish to do a phenol/chloroform extraction at this point, but it is usually not necessary.
4. Add 625 µl of ice-cold 100% ethanol and invert the tube several times to mix.
5. Freeze in an ethanol–dry ice bath for 5–15 minutes.
6. Centrifuge in a microcentrifuge for 5 minutes and carefully pour off the supernatant.
7. Remove the residual salt by adding ~1 ml of 70% ethanol, vortexing briefly and centrifuging for 5 minutes. Carefully pour off the ethanol. Repeat 2 times.
8. Dry the pellet under vacuum to remove any residual ethanol.
9. Dissolve DNA in 100 µl of TNE.
10. Analyze the DNA recovery and purity by measuring the OD_{260} and OD_{280} of a 1:100 dilution in 0.5 ml TNE.
11. Dilute the plasmid DNA to the desired concentration (usually 0.5 to 1.0 µg/µl).

REFERENCE

D.B. Clewell, and D. Helenski, *Proc. Natl. Acad. Sci.* USA **62**, 1159 (1969).

INDEX

DATE DUE